PROVING MY THEORIES

By
Lenard Metzger

Elemental Publishing
Rochester, New York

Other books by Lenard Metzger

COMMON SENSE COSMOLOGY

BEYOND EINSTEIN

THE THEORIES OF LENARD METZGER

PROVING MY THEORIES

Copyright © 2014 Lenard Metzger

Elemental Publishing
80 Westerloe Avenue
Rochester, NY 14620
(585) 473-9303

Library of Congress
Control Number: 2014935929

ID: 14338205
Lulu.com

ISBN: 978-1-304-93614-1

PROVING MY THEORIES

PREFACE

This will be the fourth and final installment of the books that I have been writing for the last ten years. The first three have the following names and dates published:

Common Sense Cosmology; 2006
Beyond Einstein; 2008
The Theories of Lenard Metzger; 2010

The present book will continue on from the last of the above books. It will describe the new ideas that have occurred to me, the corrections I have made and the actions I have taken. The first part of this book is called "Improving My Theories" and the second part is called "In Proving My Theories".

The Appendix is included at the end. It contains the chapters from my other books that are referred to within the text. Also, there is an Addendum that contains other pertinent information.

PROVING MY THEORIES

CONTENTS   PAGE

PART I. IMPROVING MY THEORIES

Although my previous books contained many different concepts (theories?) the main one that I developed, that I consider the most significant, is that of the nature of dark matter and of dark energy. I deduced that dark matter consists of neutrons, having velocities about the speed of light and that dark energy is the kinetic energy of the neutrons. These conclusions were reached in developing physical explanations for the constant speed of light and for the force of gravity (See Appendix on page 79).

Immediately after "The Theories Of Lenard Metzger" was published I realized that I had made a numerical mistake in one of the equations of the chapter entitled "Universe" (See Appendix on page 81).

The physical volume of the visible universe was calculated to be 5 X 10^{29} cubic metz years. (I had previously named the speed of light neutrons of dark matter, "metz".) Therefore, light years became metz years. Correcting the calculation, the number 5 became 14. The volume of the visible universe, in cubic metz seconds, became 14 X 10^{51}.

On the next page, this correction, when carried out, would have had one metz per second, on average, passing through every 14 square microns, instead of the 5 square microns shown previously. Since, as I had indicated, these numbers are only approximations, I decided to use an average value of 10. With this assumption, I was able to do further calculations, simply, using powers of ten. There are 10^{12} square microns in a square meter.

Therefore I have used the assumption that 10^{11} metz pass through every square meter of the visible universe, every second. These metz will come from all directions. This will be the basis of many of my future calculations.

The chapter titled Black Holes is included as reference, in the Appendix, on page 88. I have been pondering one aspect of this description of black holes. The pertinent figure and text (in italics) is copied below, from that reference, for convenience.

BLACK HOLE EVOLUTION

LEGEND

Neutrons (Up and Down Quarks)	☐
Charstrons (Charm and Strange Quarks)	☐
Tobotrons (Top and Bottom Quarks)	■

Black Hole	Galaxy Core	Galaxy	Visible Universe
10 miles	100 miles	2000 miles	10 million miles

NOT TO SCALE

In the figure above the possible evolution of black bodies is described. They range from the black body, from a single large star, all the way to a single body containing all the mass of the entire visible universe. These different bodies in this diagram are only suggestive of possible sizes and proportions.

Using the information of the metz density per square meter per second, I tried to determine how the proportions of the various zones of these bodies could be estimated. As an example take the black hole with a diameter of ten miles or 16 kilometers. The surface area will be 8×10^8 square meters. The total metz impinging on its surface will be about 8×10^{19} per second.

The chapter on Dark Energy is included as reference on page 91. In that chapter, I calculated the kinetic energy of a single metz as 7.5×10^{-11} joules. Therefore, the total energy applied to the above surface is about 6×10^9 joules/second (watts). Would this be enough energy to produce a change to the interior of the body? However, over a period of eons, the diameter of this body could accumulate sufficient neutrons to increase the diameter enough to have a surface area that receives a much greater amount of energy and produce changes in the interior.

Consider the neutron star. As stated, exotic particles are formed in the interior. At some point the concentration of energy must be sufficient to form them. Using the above calculation, with an external diameter of ten miles, the areas of spheres within this body decrease as the inverse squares of the ratios of radii. As an example:

At an internal radius of one mile, the ratio of radii is 5 to 1. The increased concentration of energy, per square meter, is 25 times greater, at the one-mile radius. As the internal radius goes towards zero the concentration of energy will go towards infinite.

Therefore, I now believe that the visible universe, in the previous figure should look something like the following.

Visible Universe Black Body

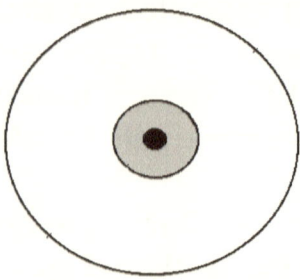

The black center sphere would consist of tobotrons. The intermediate gray sphere would consist of charstrons and the largest white sphere would be made up of neutrons. I assumed that each volume would contain about a third of the total dark matter mass. The ratio of radii, each to the next, might be about four or five to one. The diameter of the tobotron sphere could be about 7 million miles. The diameter of the charstron sphere could be about 30 million miles and the neutron sphere's diameter could be 120 million miles.

At the instant of the big bang, (See Appendix, page 85) the volume of the tobotron sphere would increase by a factor of 10,000. Its diameter would increase by a factor of 21, becoming about 150 million miles. The volume of the charstron sphere would increase by about 100 times. Its diameter would increase by a factor of about 5 and increase the combined diameter to about 300 million miles.

The neutron volume would be added to the others and the instantaneous total diameter would approach 400 million miles. Of course this is all based on assumptions upon assumptions.

As I stated in the chapter on Neutrons, (See Appendix, page 92.) the nature of the neutron is critical to my theory. The magnetic characteristics of a neutron was assumed to be as shown in the figure below, this was copied from that reference. The accompanying text is in italics.

NEUTRON

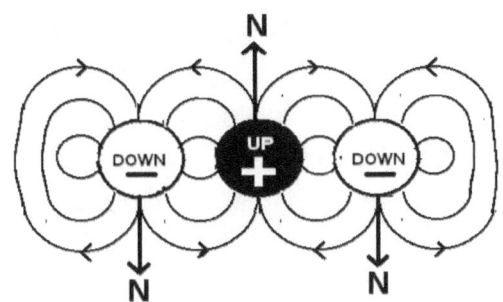

Magnetic Fields

One could lay three bar magnets side by side with the polarities arranged as those above and they will move together and hold each other firmly.

I have found by experiment that the above statement may not be entirely true. In working with toroidal magnets (more about this later) I have found that a similar configuration to the above, with the magnets side by side and with the magnetic polarities alternating, is only in a quasi-stable state.

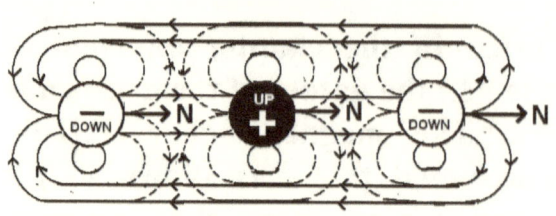

Magnetic Field

In the figure above I show what I have assumed the stable configuration of a neutron to be. I have found that toroidal magnets in contact in series, as the above, are in an extremely stable configuration.

The last chapter that I have had second thoughts about is the one titled "Proof" (See Appendix, page 94). I am including below a copy of a figure from that reference.

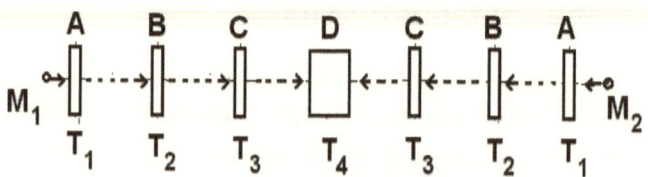

METZ EXPERIMENT

This concept assumed that pairs of metz (speed of light neutrons) would each enter opposite ends at about the same time. The various elements (A through D) would be pulsed in sequence, to interact with the metz, at the correct time intervals.

Later, I produced the following figure of a method of triggering the pulses at the proper intervals.

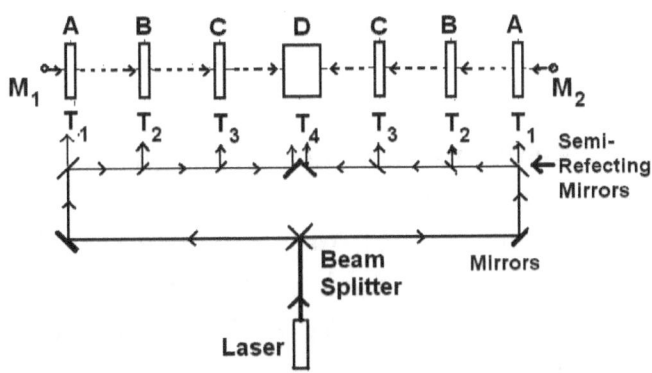

METZ EXPERIMENT 2

It occurred to me that metz would enter the device entirely at random. Leaving the elements energized continuously, instead of pulsed off half the time, would double the probability that two would enter opposite ends somewhat simultaneously.

With the new concept of the probable structure of the neutron, described previously, I thought that magnetic elements would be better than the electrostatic ones to interact with the metz. With this idea, I decided to construct a device such as the above, using permanent magnets.

13

PART II. IN PROVING MY THEORIES

First Linear Device

In researching permanent magnets, I found that neodymium toroidal magnets seemed to be what I wanted. I decided that the most powerful and cost effective size had a ¾ inch outside diameter, a ¼ inch inside diameter and a ¼ inch thickness. My first construction followed the above plan. It consisted of multiple magnets placed in series. There were 12 magnets spaced 3 ½ inches apart, all with the same polarity orientation, and covering one meter.

My assumption was that a metz approaching the input magnet, having the quasi-stable configuration, transverse to the magnet's axis, would convert to its more stable form, parallel to the magnet's axis. (See Part I, above, on the neutron.)

Metz with the stable form and having a polarity that was complementary to the magnet's polarity would be attracted and would tend towards the axis. Metz with the opposite polarity would be repelled and would tend away from the axis.

I also assumed that if metz passed through the body of the magnet, they would have their the polarity randomized by the close encounters with the magnet's atoms. Obviously, metz approaching the input magnet at the other end of the device would have all the relative polarities reversed.

I did a quick calculation of how many metz might be expected to pass through this device, each second. I assumed that the inside diameter of the end magnets would determine the number entering.

The number of metz per second reaching the second magnets came out to be about 3 thousand per second. The approach taken to obtain this number was as follows:

The aperture of the input magnet was calculated in square microns. It was assumed that for every 10 square microns one neutron entered. These neutrons come from all two-pi steradians of the sky's hemisphere. Therefore, the number of neutrons entering the aperture, from the central steradian, is reduced by a factor of two-pi.

The solid angle that the first magnet's aperture makes at the second magnet is the ratio, of the first aperture's diameter to the distance, squared. This is the ratio of ¼" to 3½", which is 1/14. This squared is 1/196. Combining this with the other reduction of two-pi makes a total reduction a factor of about 1200.

Another calculation involved converting from inches to millimeters (25.4) to microns (1000). The radius of the aperture is 3.175×10^3 microns. As usual, the area of the aperture, in square microns, is pi times radius squared. The results is about 3×10^7 square microns. This should allow for 3×10^6 neutrons per second to enter. After the attenuation factor of 1200, the neutrons entering the second magnet should be about 3000 per second.

Measurements

I had bought a Geiger counter to try to detect any particles that might have been created. I assumed that they would be protons and electrons. The following figure is from the reference on the Big Bang, on page 85.

It shows what I assumed; that neutron-to-neutron collisions produce electrons and protons.

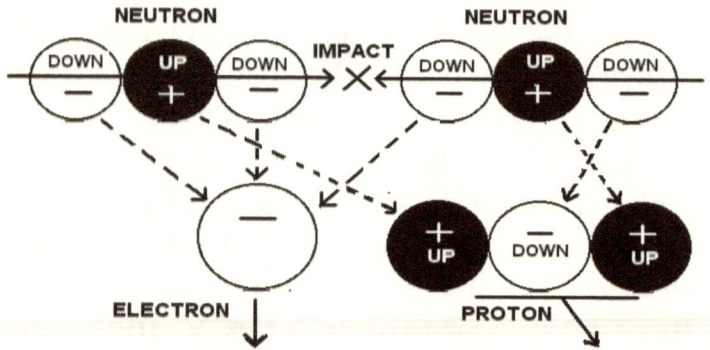

At a distance from my device, the background counts, probably mostly from cosmic rays were in the range of 10 to 20 per minute, which made it difficult to know if other particles were being produced, when the measurements were done within the device, with about the same results.

Second Device

There was also the question if 12 magnets were sufficient to collimate the metz to the magnetic axis, to allow collisions. I tripled the number of magnets to 36 and extended the assembly to 3 meters. I still could not detect any significant change from the background counts.

Below I show pictures of this device. The three sections are hinged together and held in a straight line by clasps. This third picture below is a close-up showing, with the top cap piece removed, how the magnets are spaced by slots in the wooden support. The magnets are shown being kept coaxial by having a ¼ inch diameter dowel threaded through them. At the top of the picture the cap piece is on, holding the magnets in place.

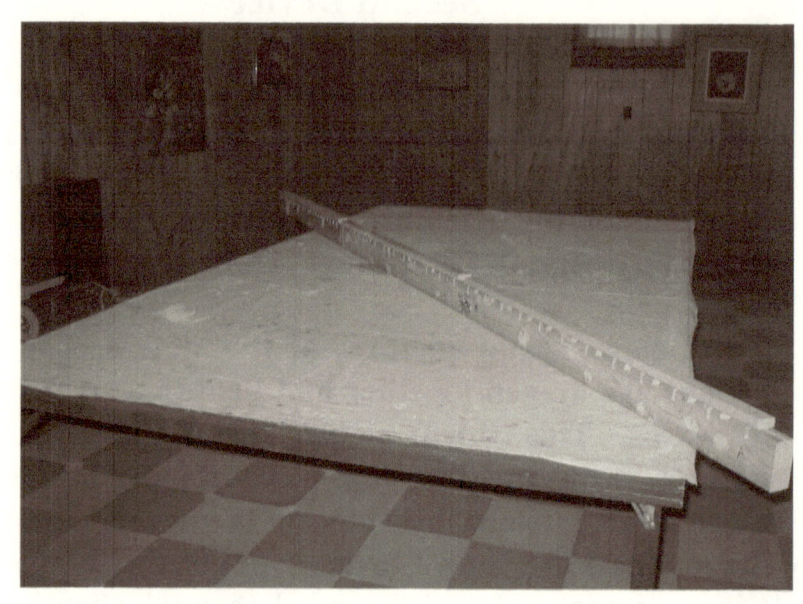

Calculations

Looking at the situation, it occurred to me that I should calculate what I could expect from this device. Being only three meters long, if neutrons traveling at the speed of light ($3X10^8$ meters per second) pass through it, they will only be in the device for a time equal to 3 meters divided by $3X10^8$ meters per seconds. This turns out to be 10^{-8} seconds.

Considering the case where two neutrons are traveling towards each other, through a device $3X10^8$ meters long, they are sure to pass each other somewhere in it. For a device only 3 meters long, 10^8 neutrons would have to pass through it each second for there to average one encounter a second.

In our case, with $3X10^3$ neutrons passing through each second, towards each other, I estimated that there would be about one encounter every 30,000 seconds. This is about one every 500 minutes or about once every 8 hours. I could have saved myself a lot of work if I had done this calculation earlier.

Parallel Path Designs

Since for me to make a much longer device was not feasible, I considered making a device with many parallel magnetic paths. Getting more neutrons per second passing through each path would also be of benefit. This latter goal could be accomplished by simply doubling the diameter of the input magnets.

My assumption was that doubling the magnets apertures would increase the number of neutrons entering by a factor of four. In addition, the solid angle that the larger aperture would present to the next magnet would decrease the attenuation by another factor of four. This change alone would increase the input by a factor of sixteen.

Many different combinations of magnets were considered for parallel path arrays. Initially, I also considered many magnets in series to address the assumed need to collimate the metz. The figure below shows the first concept that I spent some time on.

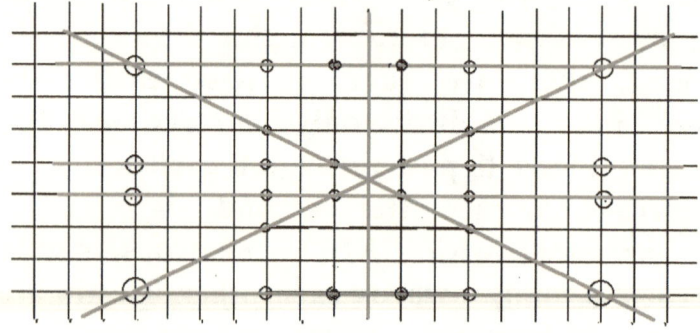

The intent was to focus the metz to the point in the center where they were supposed to collide. The sequence of magnet aperture diameters, from left to right, was ½ inch, ¼ inch, 1/8 inch, 1/8 inch, etc. There were six array holders in the design. It became apparent that six paths would not be enough. Below is another of the designs. These all ignored the problem of requiring that the metz collide at the center.

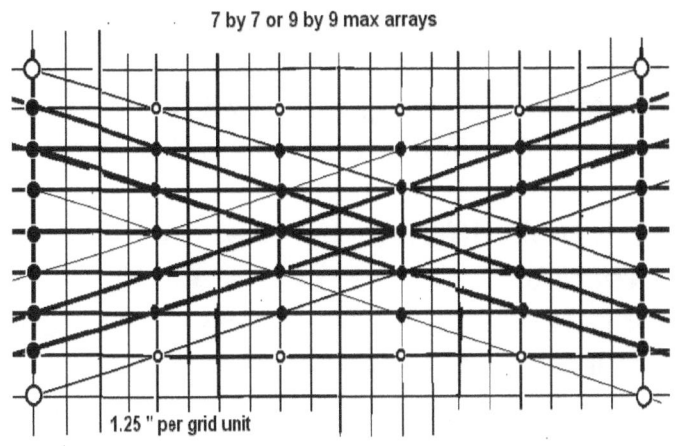

7 by 7 or 9 by 9 max arrays

1.25 " per grid unit

Two Array Design

Eventually, in considering all of the possible paths with only two eight by eight arrays, the figure shown below was produced. The concept of a simple two-array design seemed to have advantages. This figure shows some of them.

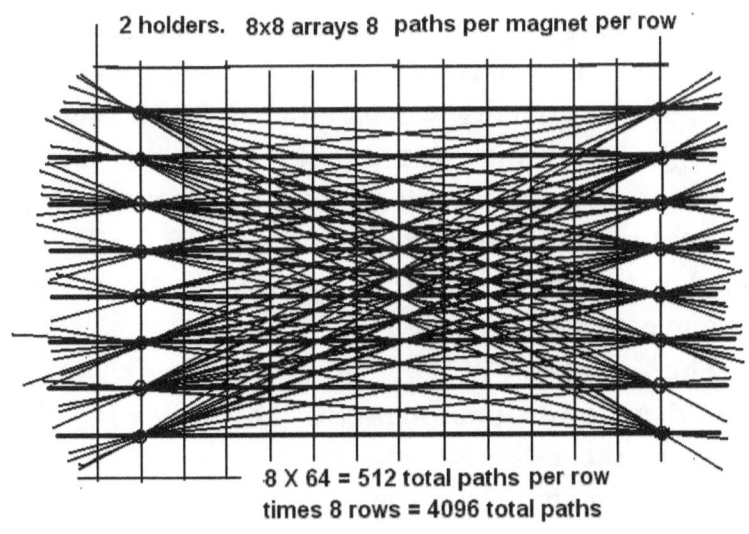

2 holders. 8x8 arrays 8 paths per magnet per row

8 X 64 = 512 total paths per row
times 8 rows = 4096 total paths

In the figure above, the vertical row of circles on the left shows one of the rows of an 8 by 8 magnet array (a total of 64 magnets). The vertical row of circles on the right, indicate the second magnet array. The significant fact is that for every magnet of each array there are 64 different paths to the other array's magnets. Therefore, there are over 4000 different paths between the two arrays. These 4000 paths, at half meter spacing, could be assumed to provide a total path length of 2000 meters.

The magnets have ½ inch inside diameters, which gives a four times increase over the ¼ inch ones. Thus the aperture area is 12×10^7 square microns. This can be assumed to have 12×10^6 neutrons per second entering the magnets.

However, at 20 inch spacing, the presented solid angle of each magnet to the corresponding one in the opposite array is $(1/40)^2$ which is a factor of reduction of 1600. Applying the two-pi reduction, the number of neutrons going from the first magnet and entering the other is reduced to about 1200 per second.

This neutron total times the distance number equals about 2.5×10^6. The ratio of this to the speed of light number predicts an encounter every 120 seconds. If the spacing were reduced to 10 inches the solid angle reduction would decrease by a factor of four, but the total length would be decreased by a factor of two. This would produce a net gain of two times. The predicted output would be increased to about one every 60 seconds.

The increase in the input magnet's aperture presented an additional benefit. The following figure shows how the effective aperture of a ¼ inch inside diameter, ¼ inch thick, top magnet, depends on the angle and distance to the next array's bottom magnet.

22

It is evident that with ¼ inch input magnets the arrays could not be spaced 10 inches apart.

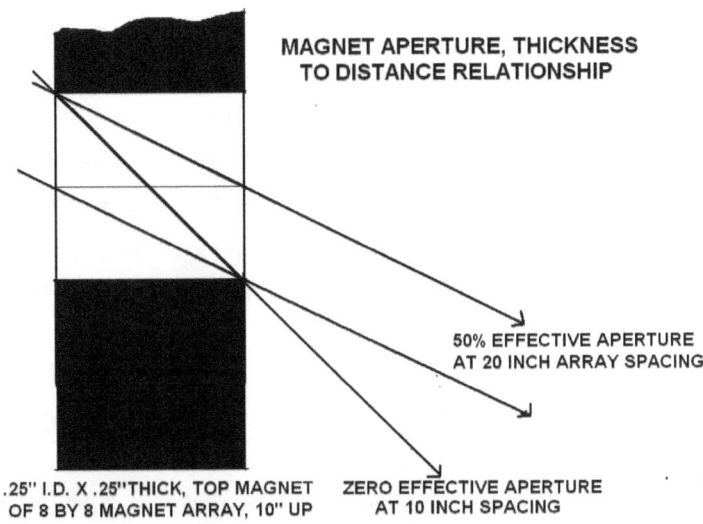

MAGNET APERTURE, THICKNESS
TO DISTANCE RELATIONSHIP

50% EFFECTIVE APERTURE
AT 20 INCH ARRAY SPACING

.25" I.D. X .25"THICK, TOP MAGNET   ZERO EFFECTIVE APERTURE
OF 8 BY 8 MAGNET ARRAY, 10" UP   AT 10 INCH SPACING

The following figure shows the same analysis for a ½ inch inside diameter input magnet. It appears that a 10 inch spacing between arrays is feasible.

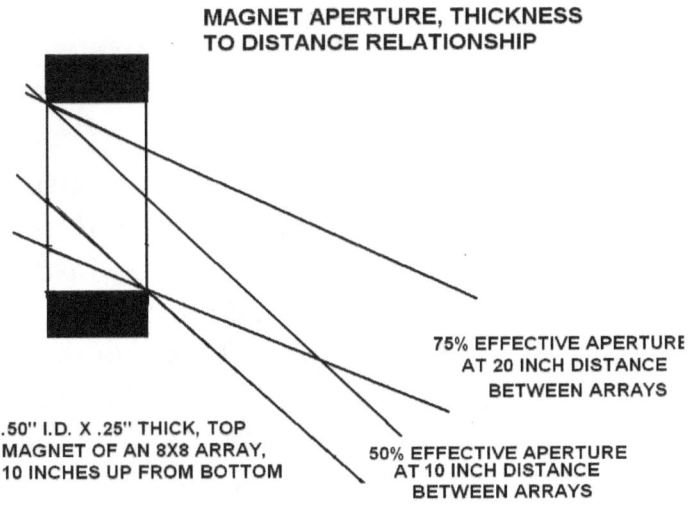

MAGNET APERTURE, THICKNESS
TO DISTANCE RELATIONSHIP

75% EFFECTIVE APERTURE
AT 20 INCH DISTANCE
BETWEEN ARRAYS

.50" I.D. X .25" THICK, TOP
MAGNET OF AN 8X8 ARRAY,
10 INCHES UP FROM BOTTOM   50% EFFECTIVE APERTURE
                                 AT 10 INCH DISTANCE
                                 BETWEEN ARRAYS

Three-Array Design

Another significant fact became evident in the earlier figure of the multiple paths of the two arrays design. Half way between the two arrays the various paths crossed in an unexpected, but obvious, pattern. As shown, the path crossings clustered vertically, in a number of nodal points, with empty spaces between them. This allowed the concept of having a third array, midway, with magnets located at the crossing points. This would increase the potential collimation of the metz.

I obtained neodymium magnets with ¾ inch outside diameters and ½ inch inside diameters and ¼ inch thickness for the two outside arrays. They were ½ the strength of the previous magnets but were less expensive. Magnets with 1/2 inch outside diameters and ¼ inch inside diameters were obtained for the third array.

Device Design

After many design attempts I settled on an approach for the device assembly using multiple arrays of magnets in flat holders that could be inserted into slots in a rectangular device. I included 5 pairs of slots, equally spaced, 5 inches apart. This enabled placing the arrays at various distances apart. The picture below shows this device. It is about 24 inches long with the ends about 12 by 12 inches square.

The picture below shows holders for the 8 by 8 arrays. The array holder on the left is fastened to its bottom cover. Its top cover is above it. The array holder on the right is without its covers. Both have ¾ inch diameter holes, ¼ inch deep.

The picture below shows the top cover on the magnet holder assembly.

To introduce a middle array holder required that the smaller magnets be twice as close together, as they were in the other two holders. Since the first two holders spaced the magnets at 1.25 inches apart, the third holder would require a magnet spacing of 5/8 inches, center to center. I selected neodymium toroidal magnets with ½ inch outside diameters and made holes to fit them.

I discovered that between the actual magnets in the arrays there seemed to be virtual magnets, with the opposite polarity to the real ones. The following figure shows what I believe is happening.

VIRTUAL MAGNETS BETWEEN TOROIDAL MAGNET ARRAY

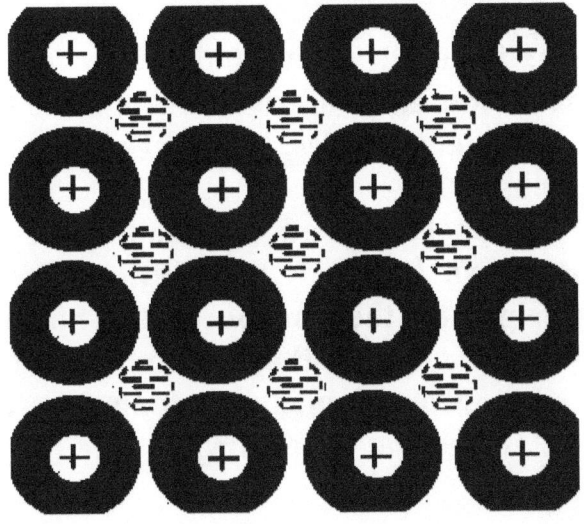

With the toroidal magnets in the array placed as close together as possible, the magnetic flux from each magnet will have to find return paths into the four larger spaces around them. Each of these spaces will return a quarter of the flux from each of the four torodial magnets adjacent to it.

The toroidal magnets have positive poles as shown, with flux emerging from them. The areas that the returning flux enters will be the equivalent of the negative poles of virtual magnets.

The significance of this is that these virtual magnets will interact with the metz of opposite polarities than those the physical magnets interact with. This has the potential of increasing the output of the device.

Array Assembling

The half-spaced array holder was very difficult to fabricate. It was easy to break through the thin walls between the magnet cavities. It was even harder to load them with magnets.

The first set-up used for loading arrays with magnets is shown below. It was difficult to maintain adequate weight on the portion of the holder loaded with magnets. At times there were eruptions, with magnets flying. Many joined into long tubes, difficult to separate. It became necessary to glue the magnets into the holder.

The first eight by eight arrays were made using up the ¼ inch inside diameter magnets from the three-meter linear device. When I decided to convert to ½ inch, inside diameter magnets I improved the assembly method to the one shown below.

In this last assembly method, an array holder was placed in a cavity. The two, 10 pound, lead weighted boards could slide over the holder and hopefully contain the magnets. It worked fine for the eight by eight arrays. With the nine by nine array and later, with the fifteen by fifteen array, there were minor eruptions, when half loaded, even after the weights were increased to 20 pounds each. The process had to be started over, with the magnets being glued in.

Three Array Device

The figure below shows the design that was built with the third holder having a nine by nine array of ½ inch magnets (81).

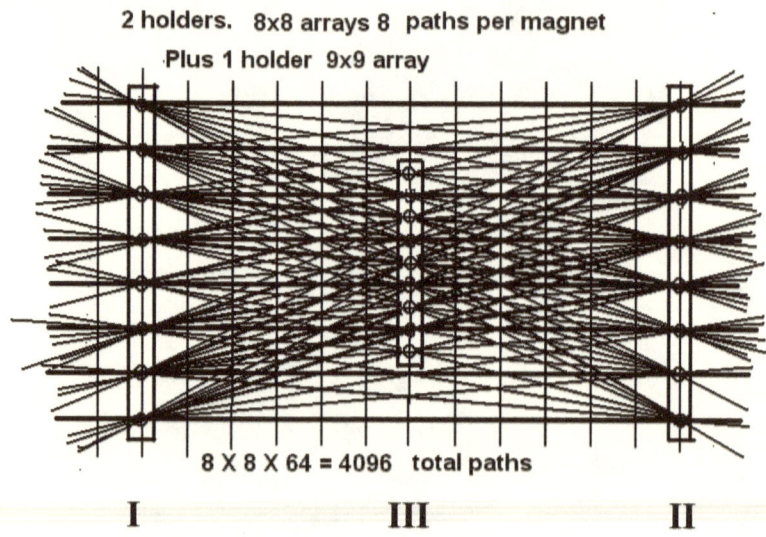

2 holders. 8x8 arrays 8 paths per magnet
Plus 1 holder 9x9 array

8 X 8 X 64 = 4096 total paths

I   III   II

I decided to limit the middle array to nine by nine because the total number of magnets increased rapidly in the outside rows of the array. Also, the number of paths crossing in the outside locations went down rapidly. It did not seem cost effective to go further at the time.

The figure below shows a nine by nine array holder, with ½ inch holes.

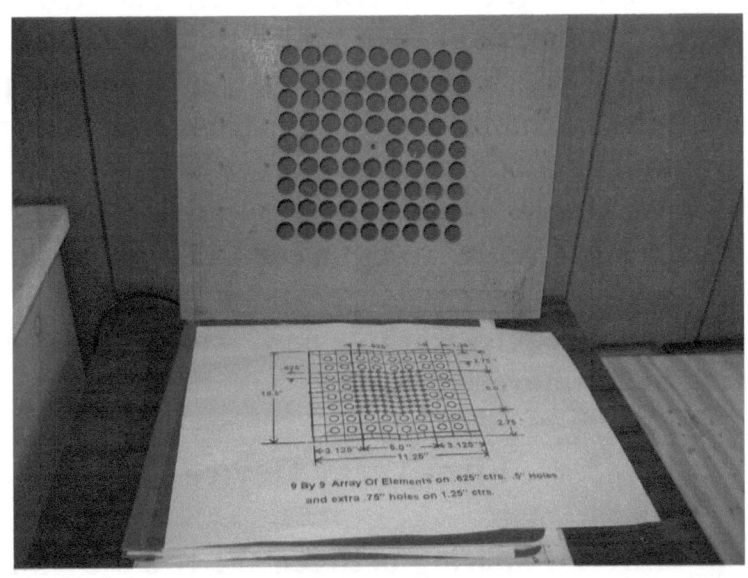

The picture below shows the three array holders in the device body, prior to the magnets being added. They are centered in the device with spare slot pairs at each end.

31

The thought of averaging only one neutron-to-neutron interaction per minute did not sound promising. The background Geiger Counter pulses were quite variable, numbering from less than ten to more than 20 a minute. The chance of proving that the device produced extra counts seemed remote. More metz or more paths were needed. To go to magnets with inside diameters larger than ½ inch, for more metz, would require a whole new, scaled up-device and would be far more expensive.

Five-Array Device

Since the spacing between the two 8 by 8 arrays was settled on to be ten inches, the other two slot pairs in the device were available. Why not add two more arrays? In the figure below a possible configuration of a device containing five arrays, is shown.

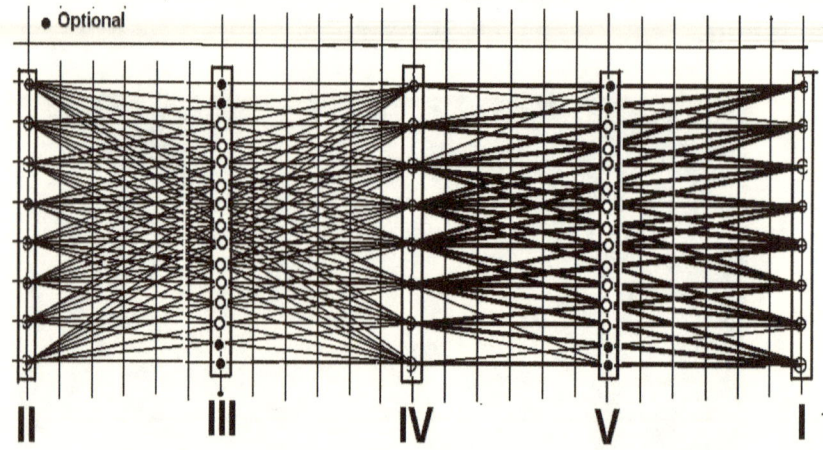

Another eight by eight array is added at the right end (I). Two arrays with closely spaced magnets (III and V) are placed mid way between each pair of eight by eight arrays (I and IV, II and IV). Arrays II and I have ½ inch inside diameter magnets.

Array IV is one of the earlier arrays tried with the more powerful ¼ inch inside diameter magnets.

All of the two dimensional paths are shown between the arrays on the left (Arrays II and IV). On the right, the only paths shown, between the arrays numbered IV and I are with dark lines. They are the ones that continue all the way to array II. The other paths are not shown.

The arrays III and V are indicated to be eleven by eleven (121 magnets each). The black dots indicate the results of increasing them to a thirteen by thirteen array (169 magnets) and a fifteen by fifteen array (225 magnets).

In the assembly of five arrays the calculation of the number of paths is as follows: The number of paths that go between all five arrays, started at any magnet of either end array, went to a four by four group of magnets (16) in the center array. These 16 paths continued to the eight by eight array at the other end. The total number of these paths is sixteen times sixty-four, or about 1000. They total about 500 meters in length.

Each of the two end assemblies of three arrays, with two eight by eight arrays, would have 4000 paths, for a total of 8000. But 1000 of these paths in each pair are accounted for above. The remaining total number of paths is about 6000. All are at least ¼ meter long, so they total 1500 meters. Therefore this gives a total length of 2000 meters.

The spacing from one array to the next is 5 inches. The ratio of the end magnet's diameter of ½ inch to 5 inches is 1/10. This squared gives a decrease of only 100 times two-pi.

The number of metz reaching the second magnet would be 12X10^6 divided by 600 or 2 X 10^4. The product of this number and the total length of 2000 meters is 40 million (metz-meters per second), a factor of 7.5 less than the speed of metz (light). Therefore, the predicted number of encounters is about one every 7.5 seconds, or 8 per minute. This might be detectible.

The actual device that was assembled required making another half space array designated V, above. I decided to make this a fifteen by fifteen magnet array. The array III was previously described, as nine by nine, with 81 half-spaced, magnets. This array also had 24 additional magnets around the periphery. The difficulties that arose in making these half spaced arrays were described above.

The picture above shows the device waiting for its last array. The fifteen by fifteen array holder is shown below, ready for it's magnets.

Data

The measurements done on the five-array device were taken in various locations in and around the device as shown below.

FIVE-ARRAY DEVICE MEASUREMENTS

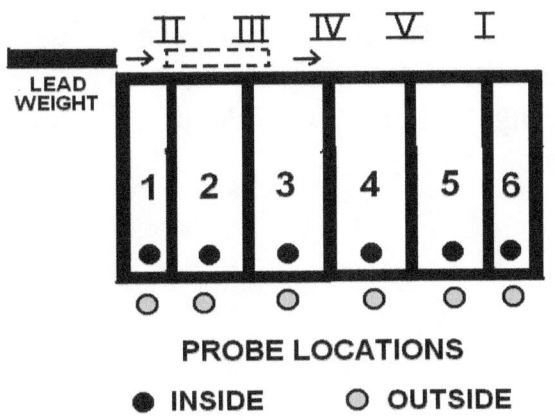

PROBE LOCATIONS

● INSIDE   ○ OUTSIDE

35

The locations of the individual arrays are designated with Roman numerals. The spaces around the arrays are designated by Arabic numerals. The locations where the Geiger probe was placed are shown as circles, black, when inside the device and gray when outside.

The first set of readings studied the effect of the lead weights over the Geiger- Counter (GC) and probe, at a distance of about 15 feet from the device.

The weights consisted of thin square lead plates, 8 inches by eight inches, each of about 2 pounds. I used a total of about twenty (40 pounds) to cover the probe. They were on a step stool and the GC was either under the weights or a few feet away.

There was also a first look at any possible difference between the horizontal and vertical orientation of the device.

The data obtained is shown below in Table 1. Each data point was a count of pulses for one minute. A sequence of 5 readings was done twice, with a time interval in between. The first four sets of readings studied the background readings, with and without lead weights over the GC probe.

The counts with the weights averaged about 10 per minute. Without the weights they averaged about 11. The readings taken alongside of the device gave about the same readings.

TABLE 1   7-22-16

One Minute Geiger Counter Measurements

Device Location	Lead Weights	GC Probe Location	Sequence	1 st	2 nd	3 rd	4 th	5 th	Avg. 5	Avg. 10 Yes	Avg. 10 No
Remote	No		a	10	9	10	13	16	11.6		
Remote	No		h	12	19	10	12	14	10.6		
Remote	Yes		b	6	8	7	11	5	7.4		
Remote	Yes		i	14	13	10	12	11	12		
										9.7	11.1
Outside	Yes	2 Outside	c	6	10	7	7	12	8.4		
Outside	Yes	5 Outside	e	11	18	20	9	13	14.2		
										11.3	
Verticle	No	2 Outside	d	16	11	10	8	3	9.6		
Verticle	No	5 Outside	f	9	17	6	12	10	10.8		
											10.2
Horizontal	Yes	2 Inside	g	10	12	8	13	12	11		

For Table 2, below, the GC probe was placed inside the device, at the bottom of each of the designated positions, between the arrays. The lead plates were placed on top of the device, over each array and probe position being used.

The background readings were taken at a distance, with the lead weights over the GC. These readings were the same as before, 10 counts per minute.

The readings from within the device averaged almost 13 counts per minute.

TABLE 2.

One Minute Geiger Counter Measurements

Device Location	Lead Weights	GC Probe Location	Sequence	1 st	2 nd	3 rd	4 th	5 th	Avg. 5	Avg. 10
Remote	Yes		a	11	14	6	17	6	10.8	
Remote	Yes		n	7	8	13	12	9	9.8	10.3
Close By		Inside								
Horizontal	Yes	1	b	8	10	16	16	18	13.6	
Horizontal	Yes	1	h	8	12	13	12	7	10.4	12
Horizontal	Yes	2	c	6	11	11	15	11	11.4	
Horizontal	Yes	2	i	12	17	16	12	20	15.4	13.4
Horizontal	Yes	3	d	8	11	10	15	13	11.4	
Horizontal	Yes	3	j	17	17	15	12	17	15.6	13.5
Horizontal	Yes	4	e	12	13	16	4	11	11.2	
Horizontal	Yes	4	k	18	13	11	12	13	13.4	12.3
Horizontal	Yes	5	f	15	12	12	12	12	12.4	
Horizontal	Yes	5	l	13	13	12	9	17	12.8	12.6
Horizontal	Yes	6	g	12	17	14	9	12	12.8	
Horizontal	Yes	6	m	16	18	12	13	7	12	12.4

Overall Avg. 12.7

The readings in Table 3 were taken with the locations of the device and that of the background readings, interchanged. The results, below, were different from those of Table 2. It may have been something due to the locations or it may be just due to the variability of the readings and the small sample sizes.

38

TABLE 3.

GC-Device Locations Reversed

One Minute Geiger Counter Measurements

Device Location	Lead Weights	GC Probe Location	Sequence	1 st	2 nd	3 rd	4 th	5 th	Avg. 5	Avg. 10
Remote	No		a	6	11	16	12	11	11.2	
Remote	Yes		n	11	11	11	18	20	14.2	12.7
Close By		Inside								
Horizontal	Yes	1	b	13	12	14	9	16	15.4	
Horizontal	Yes	1	h	9	10	7	11	14	10.2	13.0
Horizontal	Yes	2	c	13	17	15	7	10	12.4	
Horizontal	Yes	2	i	9	11	10	12	12	10.8	11.6
Horizontal	Yes	3	d	12	12	17	8	13	12.4	
Horizontal	Yes	3	j	13	14	12	10	10	11.8	12.1
Horizontal	Yes	4	e	12	9	12	13	13	11.8	
Horizontal	Yes	4	k	9	9	13	12	13	11.2	11.5
Horizontal	Yes	5	f	9	13	8	8	8	9.2	
Horizontal	Yes	5	l	9	10	15	11	13	11.6	10.4
Horizontal	Yes	6	g	14	7	17	8	16	12.4	
Horizontal	Yes	6	m	11	16	10	11	5	11	11.7

Overall Avg. 11.72

In Table 4 below, the experiment was to disable the device, supposedly, by removing a significant array. The background reading was the "out" data. As can be seen, there appeared to be a 1 count per minute difference between the device's active and inactive readings.

TABLE 4

One Minute Geiger Counter Measurements

Lead Weights	GC Probe Location	Array Number	Sequence	1 st	2 nd	3 rd	4 th	5 th	Avg. 5
			Array						
Yes	2 Inside	I	In	9	9	16	16	11	12.2
			Out	7	11	13	13	11	11
		V	In	12	11	14	11	14	12.4
			Out	9	10	9	14	9	10.2
		IV	In	10	16	8	11	13	11.6
			Out	14	13	9	12	12	12

Average In 12.1

Average Out 11.1

One other theory that could be proven with my device is that the source of weight (gravity?) is due to the imbalance of Metz coming up through the Earth relative to those coming from the sky. (See "Weight" in the Appendix on page 98).

By taking reading with the device horizontal and comparing them to readings taken with the device vertical, they might show a significant difference.

In Table 5 below, the experiment was to test the effect of the horizontal versus vertical orientation of the device. The lead weights were not used. The GC probe was fastened to the bottom of position 2, between the arrays, II and III.

TABLE 5.

7-27-13

One Minute Geiger Counter Measurements

Lead Weights	GC Probe Location	Array Location	Device Orientation	1 st	2 nd	3 rd	4 th	5 th	Avg. 5
None	2 Inside		Horizontal	14	13	15	11	9	12.4
		II Down	Vertical	11	10	16	8	9	10.8
			Horizontal	11	12	16	5	8	10.4
		II Down	Vertical	9	16	11	11	13	12
			Horizontal	19	8	13	11	14	13
		V Down	Vertical	10	12	9	8	16	11
			Horizontal	14	13	9	12	12	12
		V Down	Vertical	9	11	13	9	10	10.4

Average   Horizontal   12.5
Average   Vertical   11.1

The data in Table 5 above seemed to indicate that with the device in the vertical position the readings are close to the background readings. The horizontal position gave about the same results as the previous active readings. Table 1 had given about the same results.

40

There is an anomaly in one of the readings with the II array in the horizontal position. There was one reading of 5 counts in a minute. This caused the average for that set to be less than the vertical average. This shows how susceptible these reading are to variability.

Measuring Methods

Using the Geiger counter to obtain data was long and tedious. For most data points I did a ten-minute manual count. Earlier measurements that I made suggested that the fifteen by fifteen array might have produced a significant effect.

This could mean that the two, half-spaced, arrays might be interacting to produce additional paths. Perhaps 81 times 225 or 18,000 paths should be added. If we subtract the 8000 paths, the remaining 10,000 at ¼ meter spacing could be an additional 2500 meters. Combing this with the 2000 meters could make the counts about 20 per minute. If both of these arrays had 15 by 15 magnets the results might be even more significant, with the number of paths going between them of 225^2, or 50,000. I decided to make another 15 by 15 array and replace the 9 by 9 array with it.

Since the aperture of the Geiger probe is about ½ inch by 1 inch, only a small fraction of the possible output of the device can be detected. Each of the four volumes between arrays is 5 by 10 by 10 inches, or 500 cubic inches.

The four sides of these volumes are 5 by 10 inches or 50 square inches. That means, that at best, the Geiger counter probe will only detect one percent of whatever passes out of that area.

There are a total of sixteen such areas plus the two 10 by 10 inch ends. It remains to be seen if using this instrument the way I have been doing is going to produce definitive results.

A small cloud chamber was obtained. The picture below shows an earlier set up with it that I tried. No useful results were produced. With a 3-inch diameter and a relatively shallow active region, it was only able to detect a few vertical tracks.

The dry ice for cooling it only lasted a few days. The required, concentrated alcohol produced fumes that adversely affected me. Perhaps a large demonstration cloud chamber, with a Peltier cooler, could produce definitive results.

The basic problem that I have been having is to get data that I can rely upon to be repeatable. One last idea that I have had may be the answer to this quandary. See the Addendum on page 66.

Two 15 by 15 Arrays.

After much additional work, a second 15 by 15 array (designated VI) was made. It was substituted for the 9 by 9 array (III). The following data was taken with this set up.

FIVE-ARRAY DEVICE MEASUREMENTS

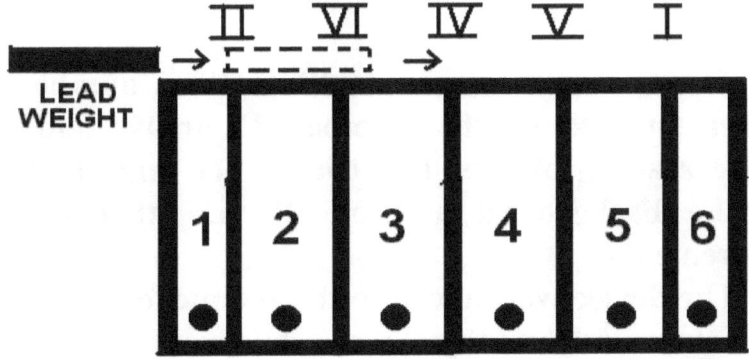

INSIDE PROBE LOCATIONS

**TWO 15 BY 15 MAGNET ARRAYS
V AND VI**

In Table 6 below, the background readings were taken before and after the device readings. In this arrangement the "remote" location was a table about 15 feet north of our regular device table. Lead weights were not used for the background readings, so they averaged about one count per minute more than when readings were taken with them.

TABLE 6

One Minute Geiger Counter Measurements

Device Location	Lead Weights	GC Probe Location	Sequence	1st	2nd	3rd	4th	5th	Avg. 5	Avg. 10
Remote	No		a	11	15	10	9	13	11.6	
Remote	No		n	14	17	14	9	8	12.4	12
Close By		Inside								
Horizontal	Yes	1	b	17	17	9	16	15	14.8	
Horizontal	Yes	1	h	13	13	17	11	16	14	14.4
Horizontal	Yes	2	c	13	15	12	8	12	12	
Horizontal	Yes	3	d	10	12	6	8	10	9.2	
Horizontal	Yes	4	e	12	10	12	17	12	12.6	
Horizontal	Yes	5	f	12	11	9	11	17	12	
Horizontal	Yes	6	g	8	14	13	9	14	11.6	
Reversed	Yes	6	m	12	9	10	14	8	10.6	11.1

These results were startling. The number 1 position, outside of the number II array, had an average reading of over two counts per minute more than the background and of all the other device measurements.

The device was moved to the remote location and the background readings were taken at the other location. Table 7 readings were taken.

Table 7

One Minute Geiger Counter Measurements

Device Location	Lead Weights	GC Probe Location	Sequence	1st	2nd	3rd	4th	5th	Avg. 5	Avg. 10
Close By	No		a	11	12	14	15	10	12.4	
Close By	No		f	10	13	8	9	14	10.8	11.6
Remote		Inside								
Horizontal	Yes	1	b	12	17	10	13	16	13.6	
Horizontal	Yes	2	c	11	11	10	7	7	9.2	
Horizontal	Yes	5	d	6	12	12	9	10	9.8	
Horizontal	Yes	6	e	11	8	12	10	13	10.8	

The same results occurred. Position 1 had readings more than two counts per minute more than the other readings.

I started investigating everything I could think of that might explain these results. Table 8 below has background readings with the Geiger counter sensor's aperture pointed in different directions to check variations. The total time of the readings was about forty minutes, which gives an indication of the variability of these readings.

TABLE 8   9-19-13

Background Geiger Counts   One Minute Measurements

Location	Vertical GC Probe Orientation	Sequence	1 st	2 nd	3 rd	4 th	5 th	Avg. 5	Avg. 10
Basement	North	a	14	11	15	12	8	12	
		b	8	9	12	20	15	12.8	12.4
	West	c	13	15	11	14	10	10.6	
		d	13	13	13	6	10	11	10.8
	South	e	11	5	10	14	16	11.2	
		f	14	13	15	9	9	12	11.6
	East	g	14	14	12	7	10	11.4	
		h	16	11	10	14	14	13	12.2

Overall Avg. 11.8

Table 8 readings were done in the usual location in the basement of my house. Table 9 data below, was taken on the first floor.

Table 9   9-20-13

Background Geiger Counts   One Minute Measurements

Location	Vertical GC Probe Orientation	Sequence	1 st	2 nd	3 rd	4 th	5 th	Avg. 5	Avg. 10
First Floor	North	a	11	12	12	12	8	11	
		b	13	11	7	13	6	10	10.5
	West	c	9	9	12	15	9	10.8	
		d	6	12	9	13	11	10.2	10.5
	South	e	14	11	9	11	11	11.2	
		f	10	8	11	11	15	11	11.1
	East	g	12	13	12	15	10	12.4	
		h	11	6	13	12	14	11.2	11.8

Overall Avg. 11.0

There is no significant difference in the data due to location or direction. The total reading time overall was about two hours and the average background count was 11.4 per minute.

Repeating the readings of Table 6. The results are shown below.

Table 10   9-25-13

One Minute Geiger Counter Measurements

Device Location	Lead Weights	GC Probe Location	Sequence	1 st	2 nd	3 rd	4 th	5 th	Avg. 5	Avg 10
Close By		Inside								
Horizontal	Yes	1	a	9	16	10	13	16	12.8	
Horizontal	Yes	1	g	9	15	17	15	18	14.8	13.8
Horizontal	Yes	2	b	14	15	9	19	9	13.2	
Horizontal	Yes	2	h	13	13	7	11	10	10.8	12
Horizontal	Yes	3	c	15	6	8	12	9	10	
Horizontal	Yes	3	i	16	11	8	13	14	12.4	11.2
Horizontal	Yes	4	d	16	12	20	11	14	14.6	
Horizontal	Yes	4	j	12	7	5	8	16	9.6	12.1
Horizontal	Yes	5	e	12	6	13	11	13	11	
Horizontal	Yes	5	k	7	8	13	12	9	9.8	10.4
Horizontal	Yes	6	f	8	15	12	10	8	10.6	
Horizontal	Yes	6	l	8	14	11	17	9	11.8	11.2

2 thru 6 Avg.   11.4

These results are essentially the same. Only the location 1 readings show the significant increase in Geiger counts, of over two more than the other readings.

I considered the possible effect of the earth's magnetic field on the magnetic fields in the device. I measured the location of magnetic north with one of the small magnets, rafted on wooden dowels, floating on water.

Magnetic north turned out to be in the north-west corner of the basement. It indicated that the red reference dot I have placed on the device magnets is the south pole of these magnets. This means that all the arrays are oriented with their south pole pointed away from the north.

Obviously, their north poles point north. The number 1 location of the device is at the south end of the magnets beside array II. This is also the end of the device away from the North Magnetic Pole.

Calculations and Analysis

The latest five array device looked like the following figure. One layer of the 1024 paths that go through all five arrays is shown as dark lines.

ARRAY NUMBERS

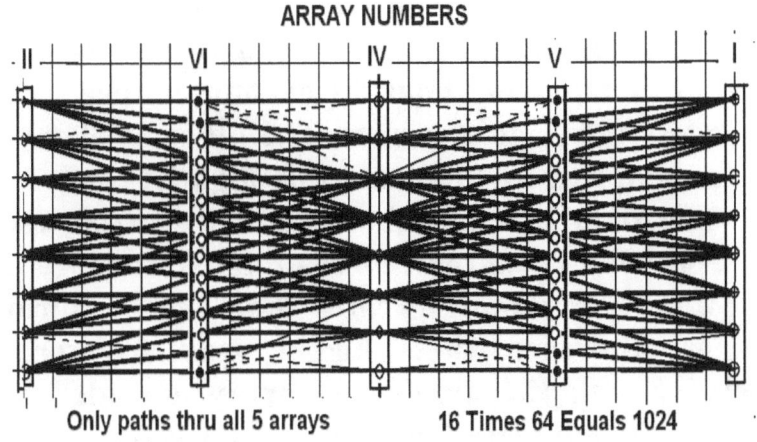

Only paths thru all 5 arrays   16 Times 64 Equals 1024

A single path through the above arrays is shown below. The vertical and horizontal scales are not equal.

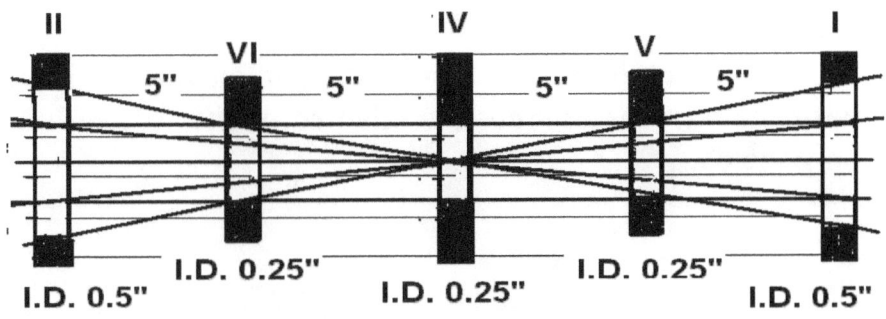

5 Holder Array, Magnet Inside Diameters Geometry

The solid angle presented by the input and output magnets (I and II) is determined by the ratio of their ½ inch inside diameters to their distance from the middle array, IV. This is ½ over 10, squared. This is $(1/20)^2$ or 1/400, which is 2.5×10^{-3}. The total number of metz per second entering, in the central steradian, is 2×10^6. Therefore, the number passing all the way through the magnets is the product of the above two terms, which is 5000 metz per second.

Each of the 1000 paths is about half a meter in length; therefore the total length is 500 meters. The path length and metz product is 2.5×10^6 metz-meters/sec. Dividing this by 3×10^8 meters/second (speed of metz) gives about .01 encounters per second. This is less than one a minute. The other 6000 shorter paths probably would increase this slightly.

The data taken is shown and discussed above. The results indicated that the readings were more than two counts per minute greater than the background readings. However, these higher readings were all associated with the number II array, outside of the device, at the left end of the device. This is the magnet's South Pole side.

My assumption, now in doubt, had been that the neutron-to-neutron collisions would occur within the device, between magnets. This is shown in the figure below.

The details of the assumed neutron interactions, with the various magnetic fields are in the Addendum on page 68.

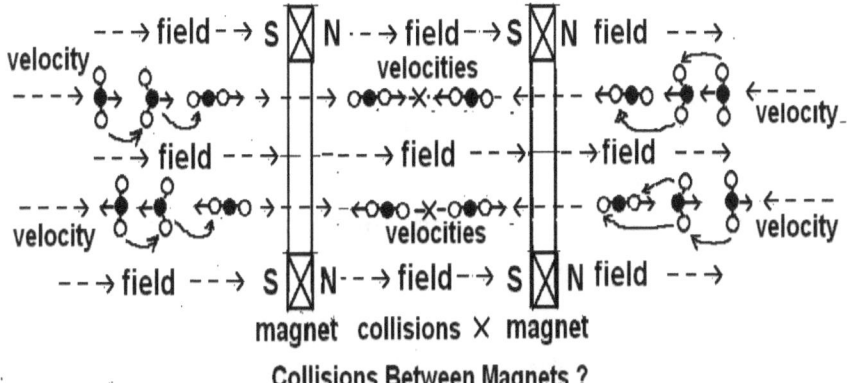

magnet collisions X magnet

Collisions Between Magnets ?

In the above figure, two neutrons are shown approaching the south pole of the left hand magnet, from the left. The upper one has its up quark with its north pole towards the magnet; therefore, its oppositely polarized down quarks will be slightly repelled by the magnet. As the neutron approaches the magnet it may convert to the linear form, having the up quark's polarity and continuing into the magnet.

The second neutron has the up quark's north pole pointed away from the magnet, with the rest of the procedure following as shown. A similar discussion applies to the two neutrons approaching the north pole of the magnet from the right.

Within the magnets, two collisions are shown between oppositely polarized linear neutrons. There are two other posibilities, with similarly polarized linear neutrons colliding.

In all cases, the collision would be between two down quarks. I have not been able to think of how the third down quark can combine with the other two and form an electron nor the other quarks form a proton.

PART III NEW THEORY

New Collisions Explanation

In the figure below, suggested by these most recent test results, the possibility of collisions external to the magnets is shown. The more complete descriptions of the assumed collisions are given in the Addendum on page 71.

Collisions External To Magnets

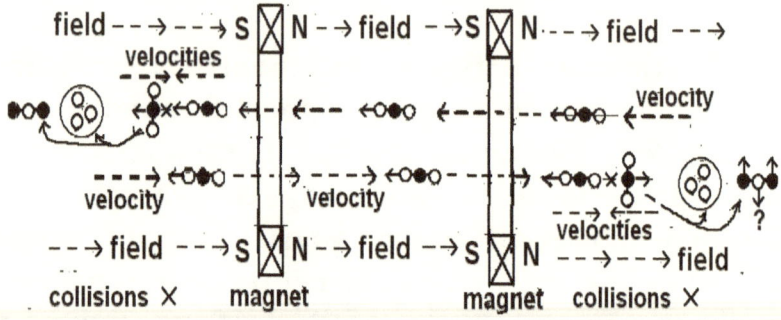

NEUTRON MAGNETIC FIELD COLLISIONS

In the above figure, I assume that the two magnets are close enough together that a continuous tube of flux passes from one magnet to the other. As a linear neutron, such as the one in the upper right of the figure, enters the magnet it will pass through and exit the left magnet.

As shown, the linear neutron collides with an approaching bipolar neutron. The bipolar neutron has its down quarks' north poles being attracted to the magnet's south pole. The collision may produce a linear proton and the three down quarks may combine into an electron.

Also shown is a linear neutron entering the left magnet and exiting at the right. It then collides with a bipolar neutron, with the north poles of its down quarks facing the north pole of the magnet. This may tend to move them away from the down quark of the linear neutron. This could make the formation of an electron more doubtful.

The polarities of the other components of the collision have different polarities and may make the formation of a linear proton impossible and that of a bipolar proton, as shown, less likely.

If this analysis is correct it may explain why the data shows most of the extra counts occurring outside of the south pole of the device.

Five Array Path Study

The following shows a detail of part of the five-array device's paths, both as inputs and outputs.

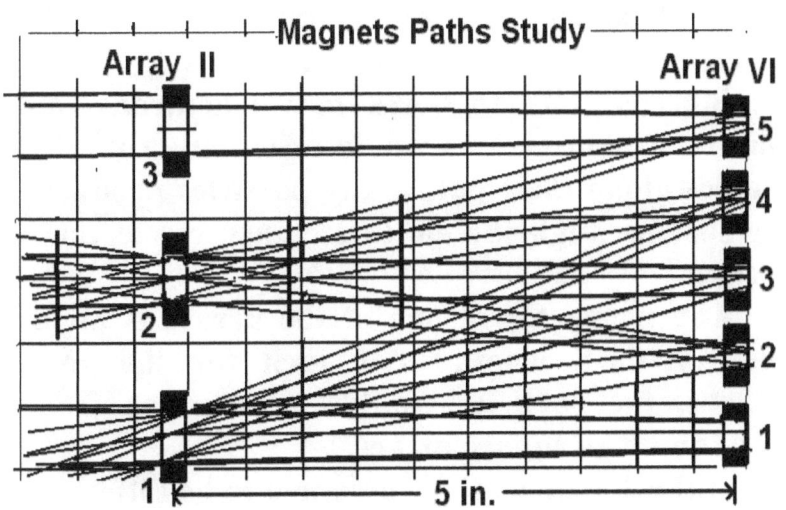

The start of a straight through path goes from magnet number 3 of array II, through magnet 5 of array VI.

Also shown, are the magnets numbered 1 and 2 of array II. Each has four paths shown, of the sixteen paths that converge into them, from array VI.

Consider II as the output array. Each of the magnets is receiving sixteen times the amount of metz calculated previously. This is 16 x 5000, or 8 x 10^4 metz per second.

Also, consider array II as the input. The central steradian number of metz, converging to about half an inch inside each magnet, is about 2 x 10^6 metz per second. The product of these two numbers, 8 x 10^4 times 2 x 10^6 equals 1.6 x 10^11 metz per second.

In the above figure, most of these metz from the many paths appear to combine in a region around a magnet, of about plus and minus one inch. This two inch length is about a twentieth of a meter.

Therefore, the metz-distance product is 1/20th of 1.6 x 10^11 or 8 x 10^9. Dividing this by the speed of metz, 3 x 10^8, gives 27 as the number of possible encounters per second in the 2 inch zone around each magnet of array II. This amounts to about 1700 encounters per second for the 64 magnets.

Since a collision between two neutrons coming from opposite directions would essentially cancel their velocities along their paths, any particles produced by the collisions might be emitted transverse to the path direction.

The perimeter of one of the arrays is about 40 inches. With a length of 2 inch for the zone of collisions, the particles would be emerging from an area of about 80 square inches.

If the 2 inch length is assumed to be different, the number of encounters calculated and the number detected each change by the same amount and compensate.

The Geiger counter probe has an aperture in its metal housing of about ½ square inch. Therefore, it would intercept a factor of 160 times less than the number of particles emerging from the 80 square inches.

If all of the above encounters produced collisions, with emitted particles, then 1700 divided by 160 or almost 11 per second, would be detected. This is 660 per minute. Since only 2 to 3 counts per minute were detected, this would indicate that only .4% of the encounters might be producing collisions. This is a much more reasonable assumption than the original one, which required every possible encounter to produce a detected count.

Array Magnets

It occurred to me that the 15 by 15 arrays could be canceling the potential benefits of the virtual magnets of the 8 by 8 arrays. Within these arrays, the virtual magnets seem to constitute a 7 by 7 array of 49 elements. Many of the 15 by 15 arrays magnets are aligned with them, with opposing polarities, which would counteract them.

The smaller apertures of the 15 by 15 arrays probably limit the number of metz passing through the various paths. Therefore, I decided to reevaluate the three array and two array configurations, using only 8 by 8 arrays.

Also, I determined the distances at which pairs of like magnets attracted each other. This would give an indication of whether a continuous tube of magnetic field could be formed between magnets, as a function of spacing.

The smaller, 1/4 inch inside diameter, magnets started to attract each other at 2 to 3 inch spacing. The larger, 1/4 inch inside diameter, magnets did so at 4 to 5 inches. I assume that the ½ inside diameter magnets would be intermediate, at 3 to 4 inches. This might indicate that a spacing of 5 inches between arrays could be spanned by a magnetic flux tube but that a 10 inch space might not be.

Array Spacing

The following 2 tables contain data taken with three 8 by 8 arrays. The first table is with 10 inch spacing between arrays. The second table is data taken with 5 inch spacing between arrays.

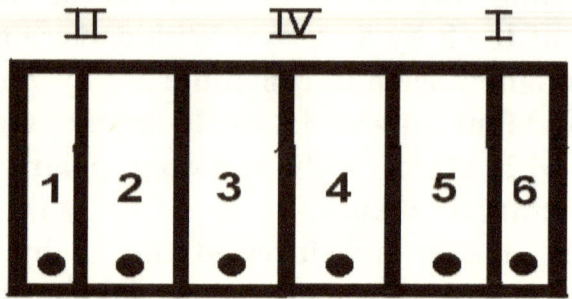

THREE -ARRAY DEVICE MEASUREMENTS

● INSIDE PROBE LOCATIONS

THREE 8 BY 8 MAGNET ARRAYS
10 INCH SPACINGS

TABLE 11

 11-10-13

One Minute Geiger Counter Measurements

GC Probe Location	Sequence	1 st	2 nd	3 rd	4 th	5 th	Avg. 5	Avg. 10
Background	After	10	14	9	8	14	11	
Inside								
1	a	9	14	12	14	6	11	
1	l	9	20	11	12	8	12	11.5
2	b	16	16	12	16	8	13.6	
2	k	19	11	15	13	13	14.2	13.9
3	c	16	15	9	11	9	12	
3	j	7	12	14	10	13	11.2	11.6
4	d	8	11	8	16	13	11.2	
4	i	9	11	15	10	7	10.4	10.8
5	e	15	16	8	16	15	14	
5	h	12	11	10	16	8	11.4	12.7
6	f	18	12	15	9	12	13.2	
6	g	16	13	11	9	16	13	13.1

Now we can see the significance of the above data. With a spacing that is probably beyond what the magnets can span, the extra Geiger counts (2 to 3) occur inside the left hand array (location 2) and both sides of the right hand array (locations 5 and 6). The center, stronger array has only background counts on both side of it.

Table 12 below has the data taken with five inch wide spaces between the three arrays. The maximum reading was outside of the right hand array.

55

THREE -ARRAY DEVICE MEASUREMENTS

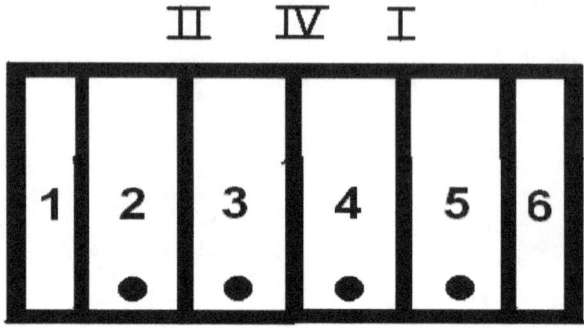

● INSIDE PROBE LOCATIONS

THREE 8 BY 8 MAGNET ARRAYS
5 INCH SPACINGS

TABLE 12

11-10-13

One Minute Geiger Counter Measurements

GC Probe Location	Sequence	1 st	2 nd	3 rd	4 th	5 th	Avg. 5	Avg. 10
Inside								
2	a	14	10	11	9	7	10.2	
2	h	10	14	13	11	12	12	11.1
3	b	11	11	11	8	14	11	
3	g	15	9	10	8	11	10.6	10.8
4	c	17	6	9	17	5	10.8	
4	f	14	12	12	14	17	13.8	12.8
5	d	16	12	13	16	11	13.6	
5	e	13	11	10	9	13	13.2	13.4

The next two tables have the data taken with only two 8 by 8 magnet arrays. For Table 13 the spacing was 5 inches.

TWO-ARRAY DEVICE MEASUREMENTS

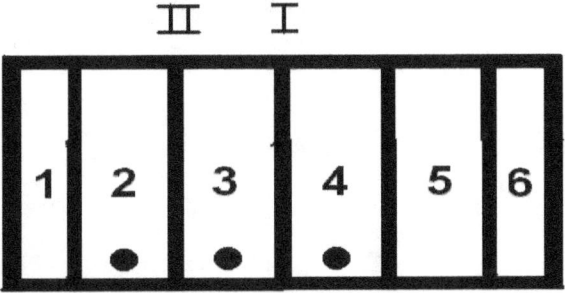

● INSIDE PROBE LOCATIONS

TWO 8 BY 8 MAGNET ARRAYS
5 INCH SPACING

TABLE 13   11-10-13

One Minute Geiger Counter Measurements

GC Probe Location	Sequence	1 st	2 nd	3 rd	4 th	5 th	Avg. 5	Avg. 10
Background	Before	10	9	6	11	11	9.4	
	After	6	16	13	11	14	12	10.7
Inside								
2		14	10	12	12	14	12.4	
3		11	17	10	5	12	11	
4		14	8	15	16	10	12.6	

The maximum readings, outside of the two arrays, were about one count per minute more than the average background count. The count between the two arrays was close to the background count.

Table 14 below is the data from the two arrays having a space between them of 10 inches.

TABLE 14   11-10-13

One Minute Geiger Counter Measurements

GC Probe Location	Sequence	1 st	2 nd	3 rd	4 th	5 th	Avg. 5	Avg. 10
Background								
	Before	8	12	10	13	11	10.8	
	After	8	13	12	5	18	11.2	11.0
Inside								
2		15	5	18	8	13	11.8	
3		13	12	12	10	13	12.0	
4		12	13	15	10	11	12.2	
5		12	15	11	15	9	12.4	
							AVG.	12.1

All the readings in and around the arrays gave a number of about one count per minute more than the background.

More Five Array Data

Since the previous five array readings apparently produced at least two counts per minute more than the background, I decided to take more readings with this configuration.

In order to be able to take readings beyond the ends of the device, the GC probe was fastened to a support and the readings were taken from the top of the arrays, with the probe aperture faced downward.

FIVE-ARRAY DEVICE MEASUREMENTS

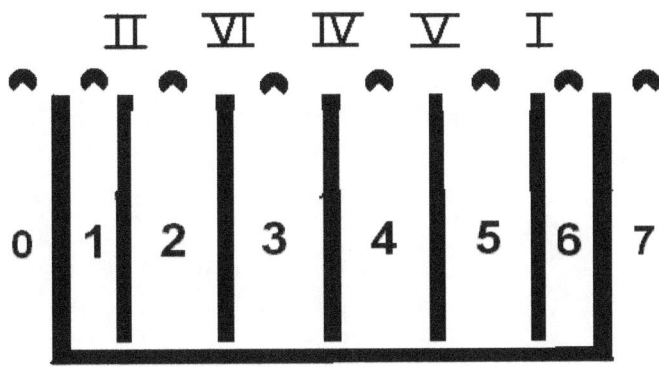

TOP PROBE LOCATIONS

LOOKING DOWNWARD

Table 15 below, has the first data with the GC probe facing downward. Each of the data, of five minute duration was taken twice. There were two additional locations used. Location "O" was just outside the left end of the device, 3 inches away from array II. Location "7" was just outside of the right end of the device, a similar distance from array I.

TABLE 15   11-12-13

One Minute Geiger Counter Measurements

Background Count Before	After	GC Probe Location	Sequence	1 st	2 nd	3 rd	4 th	5 th	Avg. 5	Avg. 10
	10.3	0	a	10	8	14	14	10	11.2	
		0	i	17	12	13	10	12	12.8	12
		1	b	13	14	15	12	8	12.4	
		1	j	12	17	14	13	14	14	13.2
		2	c	13	15	13	13	8	12.4	
		2	k	5	9	13	7	12	9.2	10.8
		3	d	13	9	14	7	6	9.8	
		3	l	9	5	8	14	8	8.8	9.3
		4	e	17	12	17	9	17	14.4	
		4	m	10	17	11	10	10	11.6	13
		5	f	19	16	4	8	13	12	
		5	n	16	9	12	14	14	13	12.5
		6	g	12	11	15	14	14	13.2	
		6	o	15	15	14	5	19	13.6	13.4
		7	h	9	15	11	17	18	14	
		7	p	10	8	9	13	13	10.6	12.3

Table 16 data below, was taken the next day, in the same manner. The background count was running slightly higher, with an average over 12 counts per minute. The intermediate locations had counts about the same as the background.

TABLE 16   11-13-13

One Minute Geiger Counter Measurements

Background Count Before	After	GC Probe Location	Sequence	1 st	2 nd	3 rd	4 th	5 th	Avg. 5	Avg. 10
11.5	12.7	0	a	19	9	14	16	10	13.6	
		0	i	12	8	9	21	19	13.8	13.7
		1	b	19	14	11	10	11	13	
		1	j	18	14	19	17	15	16.6	14.8
		2	c	17	16	7	12	10	12.4	
		2	k	10	13	15	12	17	13.4	12.9
		3	d	13	14	10	14	6	11.4	
		3	l	18	14	20	13	14	15.8	13.6
		4	e	10	19	13	16	9	13.4	
		4	m	7	7	10	10	20	10.8	12.1
		5	f	13	6	11	17	17	12.8	
		5	n	15	8	16	11	11	12.2	12.5
		6	g	19	7	9	11	11	10.8	
		6	o	9	16	18	17	15	15	12.9
		7	h	18	11	12	14	18	14.6	
		7	p	14	15	14	15	11	13.8	14.2

The next day, Table 17 data was obtained. Only one set of five minute readings were taken. There seem to be some aberrant results. I postulate that there was a period of unusual background radiation. The last data readings were very high. The last background readings can be seen to be out of the usual range of background readings.

TABLE 17   11-14-13

One Minute Geiger Counter Measurements

Time	Background Count	GC Probe Location	Sequence	1 st	2 nd	3 rd	4 th	5 th	Avg. 5
11:00 AM	Before	Remote		8	13	13	11	16	12.2
11:55 AM	After	Remote		13	25	19	14	16	17.4 *
			0	a	14	12	21	11	11	13.8
			1	b	18	12	6	14	9	11.8
			2	c	13	11	12	8	11	11.0
			3	d	13	9	18	14	15	13.8
			4	m	15	22	8	13	15	14.6
			5	n	14	8	11	18	12	12.6
			6	o	14	20	18	16	19	17.4 *
			7	h	17	15	10	11	18	14.2

Table 18 is a summary of all the previous 5 array data. Each datum is an average of 5 minutes of counts. The data of Table 17 is included in Table 18, below.

61

TABLE 18

			Background Count		Five Minute Average Count							
Source	Date	Time	Before	After	0	1	2	3	4	5	6	7
Table 6	9-11-13		11.6			13.6	12.0	9.2	12.6	12.0	12.6	
						14.0					10.6	
Table 7	9-13-13		12.4	10.8		13.6	9.2			9.8	10.8	
Table 10	9-25-13	4:15 PM	11.4			12.8	13.2	10.0	14.6	11.0	10.6	
		5:00 PM				14.8	10.8	12.4	9.6	9.8	11.8	
Table 15	11-12-13	11:00 AM		10.6	11.2	12.4	12.4	9.8	14.4	12.0	13.2	14.0
		12:20 PM			12.0	13.2	10.8	9.2	13.0	12.5	13.4	12.3
Table 16	11-13-13	4:30 PM	11.5		13.6	13.0	12.4	11.4	13.4	12.8	10.8	14.6
		6:10 PM		12.7	13.8	16.6	13.4	15.8	10.8	12.2	15.0	13.8
Table 17	11-14-13	11:15	12.2		13.8	11.8	11.0	13.8	14.6	12.6	17.4	14.2
		12:10 PM		17.4*								
Overall Averages			11.9	13.3 / 11.4*	13.0	13.6	11.7	11.9	13.0	12.5	14.0	13.8
Total Minutes			25	25	25	50	45	40	40	45	50	25

In the above, when using the 17.4 background count reading, the average background was raised to 13.3 counts per minute. Omitting it gave the usual reading of 11.4 counts.

In Table 19 below, without the data of Table 17, the background counts were in the usual range of 11.4 to 11.7 counts. The maximum number of counts was from location 1, near array II and location 7, outside of array I. They were counts of almost 14 per minute. This was about two counts per minute more than the background. Most of the final overall averages represented about forty minutes of readings, each.

TABLE 19

Source	Date	Time	Background Count		Five Minute Average Count							
			Before	After	0	1	2	3	4	5	6	7
Table 6	9-11-13		11.6			14.8	12.0	9.2	12.6	12.0	11.6	
						14.0					10.6	
Table 7	9-13-13		12.4	10.8		13.6	9.2			9.8	10.8	
Table 10	9-25-13	4:15 PM	11.4			12.8	13.2	10.0	14.6	11.0	10.6	
		5:00 PM				14.8	10.8	12.4	9.6	9.8	11.8	
Table 15	11-12-13	11:00 AM		10.6	11.2	12.4	12.4	9.8	14.4	12.0	13.2	14.0
		12:20 PM			12.0	13.2	10.8	9.2	13.0	12.5	13.4	12.3
Table 16	11-13-13	4:30 PM	11.5		13.6	13.0	12.4	11.4	13.4	12.8	10.8	14.6
		6:10 PM		12.7	13.8	16.6	13.4	15.8	10.8	12.2	15.0	13.8
	Overall Averages		11.7	11.4	12.7	13.9	11.8	11.1	12.6	11.5	12.0	13.7
	Total Minutes		20	20	20	45	40	35	35	40	45	**20**

It is apparent from the above, that with a large enough sample, a few aberrant readings will not change the results significantly.

Part IV. Conclusions

The results are clear. The averages of GC counts at the two ends of the apparatus are at least two counts per minute greater that the background counts.

It remains to be determined if the Geiger counts, in addition to those caused by Cosmic rays, are due to electrons and protons that have been created by collisions between speed of light neutrons, i.e. metz. There could also be other particles created by such collisions. There could be some source of these extra counts, other than collisions.

Further testing of my device should be done with other kinds of instrumentation that I don't presently have. The first step might be to try a larger cloud chamber with a strong magnetic field, transverse to it. If curved paths could be observed perhaps the kinds of particles produced could be identified.

Smaller versions of the add-on devices, described in the Addendum on page 66, could be made and placed at the two ends of the array holder device. The possible outputs of these add-ons can be recalculated based on the later data of my New Theory.

If all of the above verifies that my device is producing neutron-to-neutron collisions, and that other particles are being produced, then my theory of dark matter may be considered proven.

With the above established my theory of gravity may then be considered seriously and that the speed of light may be seen as caused by dark matter.

ADDENDUM

Add-Ons

Below a diagram of what I call an add-on to the device is shown. This development is to be added to my device to detect any electrons and protons that are produced.

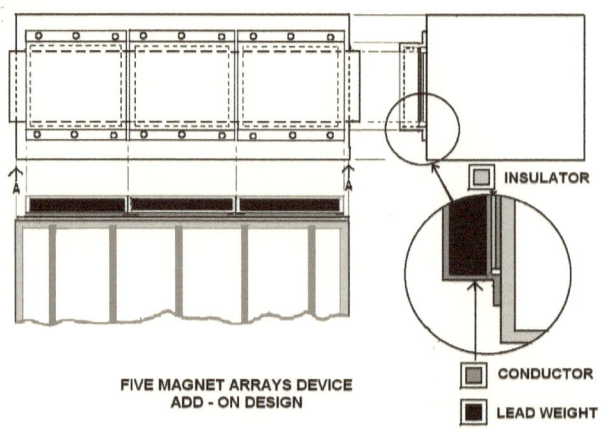

FIVE MAGNET ARRAYS DEVICE
ADD - ON DESIGN

INSULATOR

CONDUCTOR

LEAD WEIGHT

The theory behind the add-on is that if electrons are produced, the first, thin, insulated conductors will collect them. If protons are produced they will probably pass through the thin conductor and insulator and be collected by the lead weights. The protons would then transfer to the conducting metal surrounding the lead weights.

This will be, in an affect, a large capacitor. Each plate will have an output terminal. A voltage may build up between them, a negative voltage, from electrons, on one plate and a positive voltage, from protons on the other. This will indicate that the device is working.

A stack of 8 inch by 8-inch lead plates will be contained in each of the metal holders shown in the figure. There will be three holders for each side of the device and one at each end, if desired.

It is conceivable that if this device were scaled up tremendously, a significant amount of power could be produced.

To get an idea of what the present output might be, if all twenty collisions a minute produce electrons and neutrons, calculate how long it will take to accumulate enough charge to be measurable.

A coulomb is about 6×10^{18} electrons. The relationship between a quantity of electric charge (Q) and the related capacitance (C) and voltage (V) is $Q=CV$ or $V=Q/C$.

If the add-on could have a minimal capacitance of 10^{-11} farads (10 pico-farads) and if we could detect 10^{-3} volts, then Q would equal 10^{-14} coulomb, or 10,000 electrons. Assuming that we could detect and accumulate electrons at a rate of 20 per minute, it would take 500 minutes to reach a voltage of 10^{-3} volts.

At the present time I have 22 lead plates, each of them 2 pounds. I could make three add-ons each with 7 plates and add them to one side of the device. If I planned on adding these to two sides, I would need six add-ons. I could have one side with 4 plates in each add-on and the other side with 3 in each. This could give an indication of whether protons need more or less lead thickness to be stopped.

With the "New Theory" the above would be changed to the add-ons being placed around the two end arrays. The calculation of the charging time might indicate a more rapid voltage build up.

Neutron-Magnet Interactions

In the following figures I show my assumptions of how neutrons may interact with the magnets in the device. They may also have an interaction with the Earth's magnetic field.

I assume that bipolar neutrons are those with the positive quark's north pole pointed "up" and the other two negative quarks with their north poles pointed "down". The possible neutron configuration, with all the quarks having their north poles in the same direction, I call linear. The figures below show many of these possibilities.

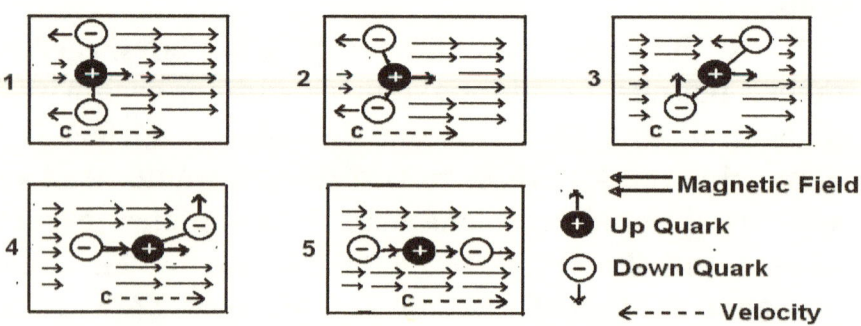

NEUTRON MAGNETIC FIELD INTERACTION 1

Figure 1 above, shows the neutron moving to the right in the same direction as the magnetic field. This may be towards the south pole of a magnet. The polarity of the up quark continues on as the neutron converts to the linear configuration.

Figure 2 below, is similar to Figure 1, except that the polarity of the up quark is reversed.

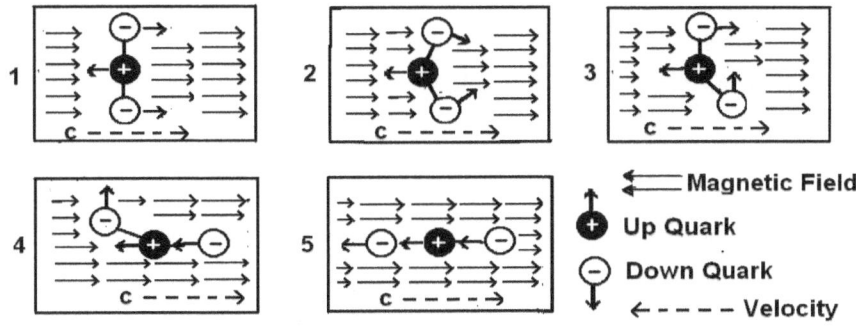

NEUTRON - MAGNETIC FIELD - INTERACTION 2

In the next two figures the neutrons are moving to the right against the flow of the magnetic field. This could be that the north pole of a magnet is to the right.

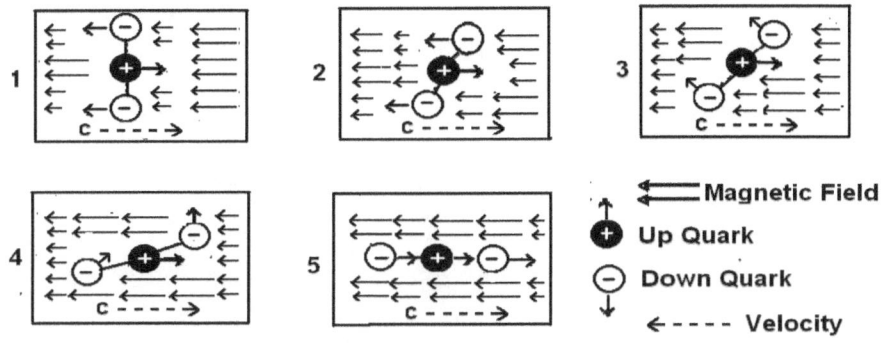

NEUTRON MAGNETIC FIELD INTERACTION 3

The up quark's north pole is directed to the right in the figure above.

In the figure below, the north pole of the up quark, is directed to the left.

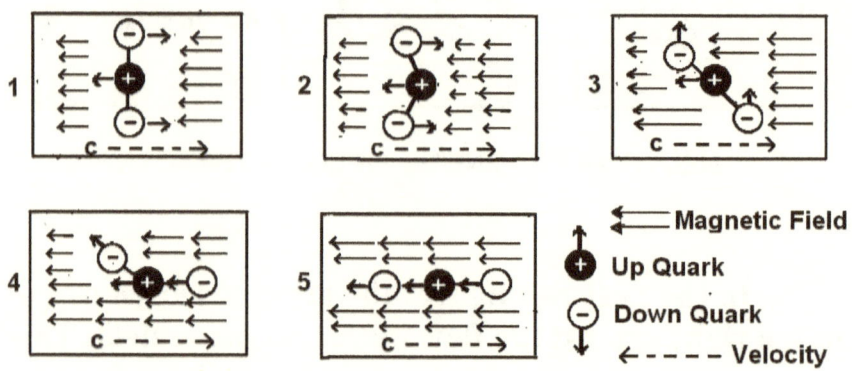

NEUTRON - MAGNETIC FIELD - INTERACTION 4

Neutron-To-Neutron Collisions

My assumption for the external collisions is that an emerging linear neutron collides with a bipolar neuron, as it is most susceptible to being split. In the figures below I show what I think are four possible encounters. Each shows four sequential configurations each ending with the resulting proton and electron.

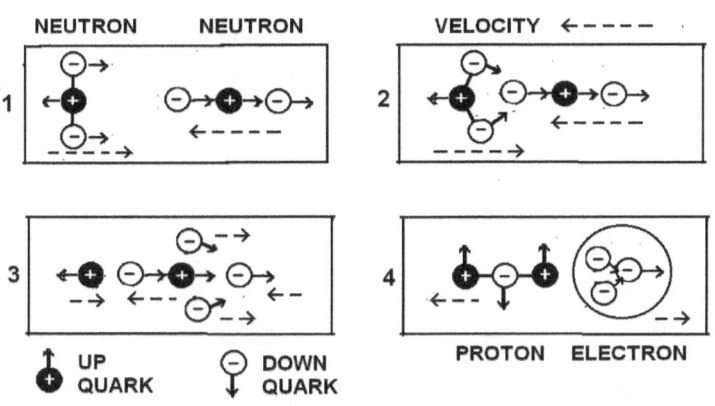

NEUTRON TO NEUTRON COLLISION 1

This collision ends with the up quarks oppositely polarized and the down quarks with the same polarization. I feel that this is unlikely to produce a proton and electron.

71

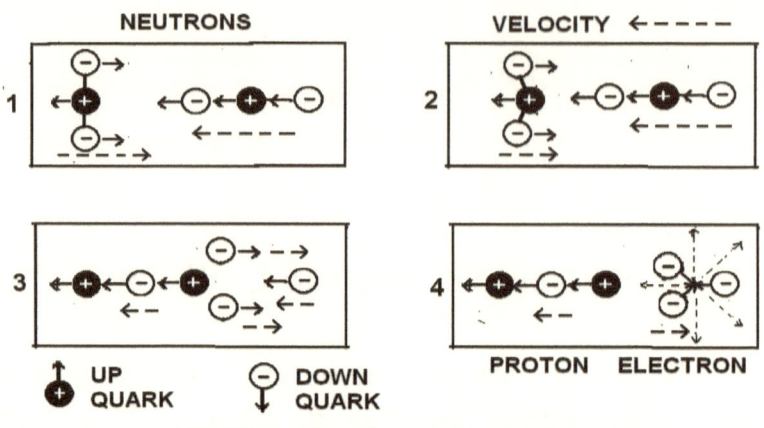

NEUTRON TO NEUTRON COLLISION 2

The above collision seems to produce quark polarities that are more likely to produce a proton and an electron.

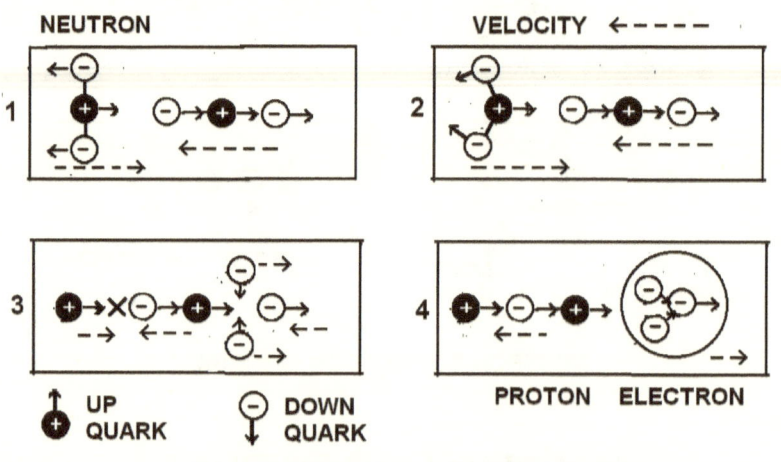

NEUTRON TO NEUTRON COLLISION 3

Collision type 3 above is also likely to produce protons and electrons. However, collision type 4, below, has the same problem as in collision 1, above.

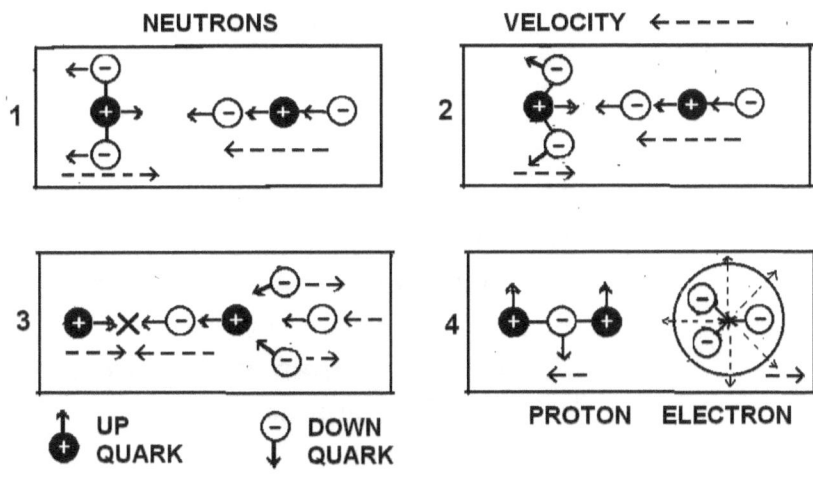

UP QUARK

DOWN QUARK

PROTON ELECTRON

NEUTRON TO NEUTRON COLLISION 4

73

Redesign

If I had it to do it over what would I do differently?

1. Make the device lighter
2. Have it made professionally
3. Use non magnetic metal for fabrication
4. Scale it up for more counts
   a. Larger diameter magnets
   b. More magnets per array
   c. Longer device
   d. All of the above

In my early designs of the structure of the device, I first did them with the thought that they would be made professionally. The figures below show some of the differences from what I designed later to be made by myself.

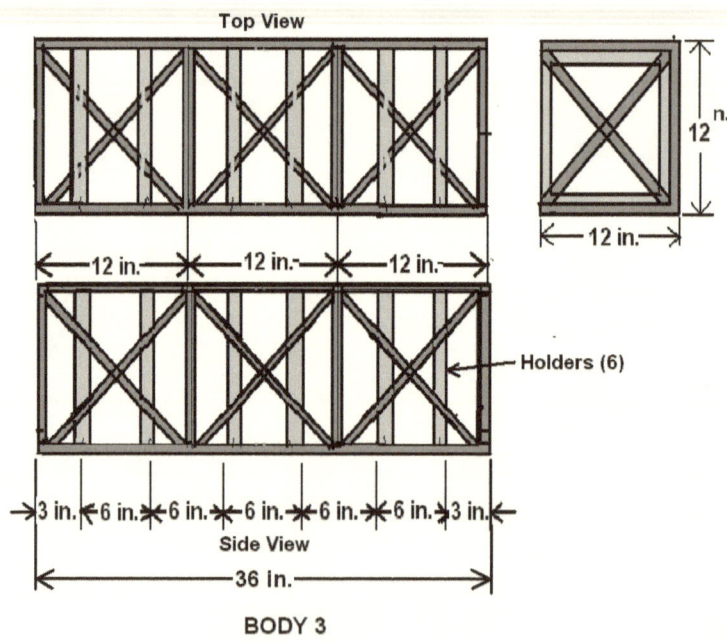

BODY 3

This is an open framework body, potentially much lighter than the present one. I assumed that it would be made from a non-magnetic material, such as aluminum. It was designed for the six array designs that I was considering at that time.

The following figure shows one-piece magnet holder designs that would be made by machining. They included axial, through-holes to allow aligning the multiple arrays optically.

Holder A and Cover   Holder B and Cover

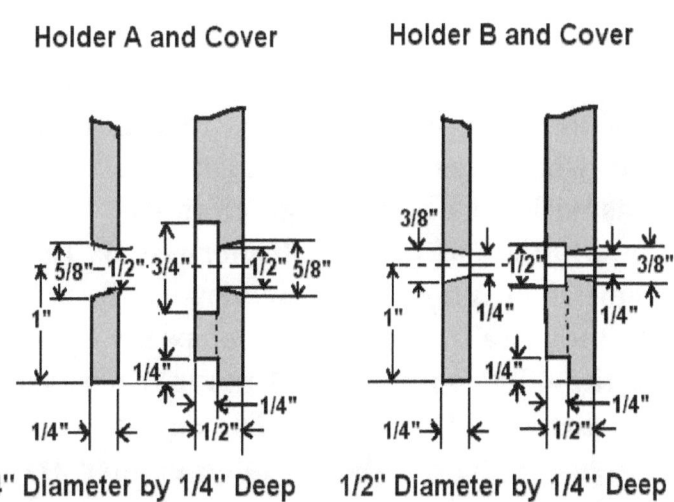

3/4" Diameter by 1/4" Deep   1/2" Diameter by 1/4" Deep

I have considered two simple methods for scaling the device up, to provide increased output counts. The calculations were based on the original theory.

The first is to double the inside diameters of the magnets. This would increase the number of input metz by a factor of four. I assume that the overall length of the device would also be doubled.

The input magnet's, inside diameter to spacing ratio will remain the same, but the increased length will improve the output directly. I expect that the total increase could show an eight-fold increase, since the number of paths will not have changed.

There would be three 8 by 8 arrays. The outer array magnets could be ones with 2 inch outside diameters and one inch inside diameters. They are ¼ inch thick. A 2½ inch center-to-center spacing would require a holder of about 21 inches on a side. The two 15 by 15 arrays could make use of the ½ inch inside diameter magnets from the present arrays, plus more.

The second approach is to keep the same size magnets, as in the present design, but increase the number of magnets in each array. My thought was to determine what was required to produce the same increase as with the above design change.

The number of paths will increase as the square of the number of magnets in the arrays. To achieve an eight-fold increase in the number of paths for two 8 by 8 arrays, the 4000 paths, for 64 magnet arrays, would have to be increased to 32,000. The square root of this number gives us the number of magnets per array, 179 magnets. This would require a 13.37 x 13.37 array. (Obviously impossible.)

Using 13 by 13 arrays gives a seven times increase. With 14 by 14 arrays (196 magnets) the increase is to almost 10 times. The half-spaced arrays would be 27 by 27 which totals 729 magnets each. The array holder for the larger case would be about 18 inches on a side.

If the length of the device is increased to 36 inches and the spacing between arrays is increased to 7 inches from 5 inches, the change in the aperture to spacing ratio will produce a factor of 2 times reduction in the number of metz per path. The increased length of the device will produce a 50% increase in total path length. The net overall output will go down from a 10 times increase to an 8 times increase. (The same as in the first case.)

With the "New Theory" idea the above calculation would change. The path lengths would not be as significant a parameter since the primary areas of metz interactions is around the end arrays.

The increase in output should go up directly with end magnets' apertures. The output as a function of the number of magnets in the arrays should increase in a different manner. The 8 by 8 array gives a multiplying factor of 16. Using 10 by 10 arrays would give a multiplying factor of 25. Also, 12 by 12 arrays would give a factor of 36.

The 14 by 14 arrays (in the previous calculations) would now give a multiplying factor of 49, which is only a three times increase over 16. This suggests that the increased magnet aperture alternative would be preferred.

I would include add-on devices at both ends of the device. These were described in the Addendum on page 66. The electron and proton accumulation should be considerably faster that the previous calculations.

APPENDIX

(Mostly Chapters From
"The Theories Of Lenard Metzger")

INTRODUCTION

I believe my most important ideas are in astrophysics. They began with my desire to reconcile the dichotomy between the electro-magnetic and the quantum theories of light. My assumption about the nature of dark matter, to explain the speed of light, was key. This concept of dark matter led me to a new explanation for gravity. This also led to explanations for black holes and the big bang.

Dark matter is believed by science to make up most of the mass of the universe. I proposed that dark matter consists of neutrons traveling at or near the speed of light. I suggested that they actually propel photons to that speed.

In another important theory, I describe how the effect of gravity is caused by the weak interaction of these particles with matter. In my previous books I had called these speed-of-light neutrons, "gravitons" to distinguish them from the familiar, low velocity neutrons usually detected. However, the term "gravitons" has a different connotation in physics so I have taken advantage of my literary license and given them a new name. With no false modesty, I will name the speed-of-light neutrons "metz " after myself. Hereafter, the speed of light will be known as the speed of metz.

In my theory of how black holes are formed I proposed that in addition to neutrons, as in neutron stars, other forms of neutral particles are created in the center of larger black bodies.

The collapse of a large star into a black hole will stop at a finite volume with particles, made of charm and strange quarks, forming in the center. I named these particles, "charstrons". Each would have the mass equivalent to about a hundred neutrons.

I also proposed that a third kind of neutral particles would be made of top and bottom quarks and would have the mass-energy of about a hundred charstrons. They would form in the center of much larger black bodies, such as the one at the center of our galaxy. I named these particles, "tobotrons".

UNIVERSE

The following figure uses a method of showing my concept of the universe.

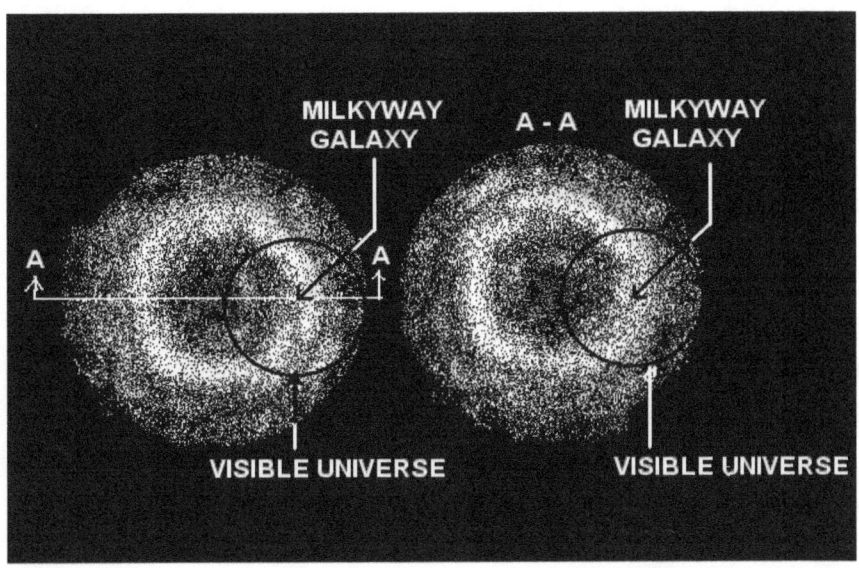

The circle displayed on the left is a cross sectional view of the universe, in the plane of the paper. The section A-A at the right is a cross sectional view at right angles to the plane of the paper. Assuming that the age of the universe is 13 billion years, the diameter (roughly from A to A) would be about 26 billion light (metz) years. As I have previously explained, I reject the relativistic concept of expanding space.

As indicated by the smaller black circles, the visible universe would extend from the center of the universe to its outer reaches. This circle has a radius of about 6.5 billion metz years. The Milky Way galaxy is assumed to be near the center of this volume.

The density of matter along the radial direction is peaked near the center of the visible universe with a some-what- normal distribution. The distribution is more uniform along the circumference at a constant radius.

The Milky Way galaxy was assumed to contain 100 billion solar masses. The visible universe black body could have the mass of 100 billion galaxies. Therefore, the visible universe black body can be estimated as having 10 to the power of 22 (10^{22}) solar masses, at a minimum.

I assumed that what was to become dark matter increased the mass of the visible universe black body by a factor of eight. This doubled the diameter of the body to 20 million miles. The visible universe accounted for one eighth of this and the rest became dark matter. I assumed the dark matter, in the visible universe, approximated 10^{23} solar masses.

Using the mass of the sun (one solar mass) and the mass of a neutron, I calculated that a minimum of 10^{80} neutrons (metz) constitute the dark matter in the visible universe.

Referring to the above figure, the diameter of the full universe was assumed to be twice that of the visible universe. Therefore, the full universe would have a volume eight times that of the visible universe.

The original diameter of the full universe size black body would be about 40 million miles. At the big bang it would expand, almost instantaneously, to about a billion miles in diameter.

In calculating the volume of the visible universe, an average radius of about 5 billion metz years was assumed.

Using the equation for volume of a sphere, 4/3 (pi) R^3, the visible universe comes out a sphere with a volume of about 5 X 10^29 cubic metz years. A cubic metz year is an awkward unit to use for this purpose. Instead I repeated the calculation in a unit I believe is more useful, a cubic metz second.

The number of seconds in a year is about 3 X 10^7. Therefore, for the visible universe, I used a radius of 15 X 10^16 metz seconds.

This gives a volume of about 5 X 10^51 cubic metz seconds. A metz second is 3 X 10^8 meters, which is 3 X 10^14 microns. Each face of a cubic metz second is about 10^17 square meters and also 10^29 square microns. This is shown in the figure below.

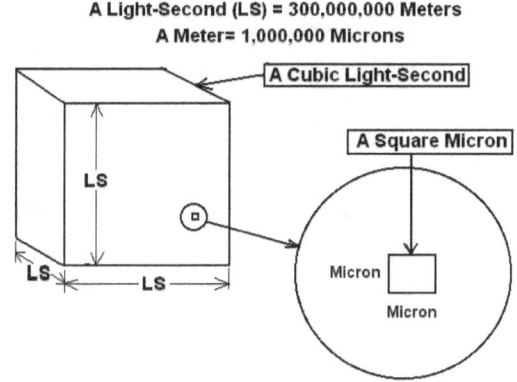

A Light-Second (LS) = 300,000,000 Meters
A Meter= 1,000,000 Microns

A Cubic Light-Second

A Square Micron

LS

LS

LS

Micron

Micron

Metz entering the above cubic metz second, normal to the face will pass through all 3 X 10^14 cubic microns in the column, in one second.

It would take a metz entering each of the square microns on the face, 10^29 of them, to pass through every cubic micron in this cubic metz second, every second.

For all the cubic metz seconds, throughout the entire volume of the universe, to be visited by metz every second, would take about 5 X 10^80 metz. This requirement is a factor of 5 more than the number of metz computed to be in dark matter, but this is within the margin of error for this calculation.

The available number of metz would appear to be enough if we set the dimensions of the elemental cube to about 2 microns. It is obvious that all metz will not be oriented perpendicular to every cubic surface at the same instant, as assumed above. The random orientations and timing of the passage through space of every metz, will make their visit to each atom highly variable.

It remains to be determined how close a metz has to come to an atom to trigger the emission of a photon or how close to a photon to accelerate it to its characteristic velocity. It also depends on how close a metz has to come to an atom of a body to impart a bit of momentum to the atom, through electro magnetic interaction, causing deflection of the atom and the metz, thus producing gravitational effects.

BIG BANG

There are several possible explanations for what triggered the big bang. An impact with another massive black body could have cracked it open. It could have become too massive and collapsed. However, I favor the concept that when the flux of metz, impacting on its surface, fell below a certain limit, the internal pressure opened up the outer shells and the explosion took place.

The following figure shows how the big bang might have looked, in a scaled down fashion. The idea was inspired by my recollection of a particular finale of a fireworks display. A single bomb was fired upward. It burst into a sphere of bomblets. Each of the bomblets burst into a large number of smaller bomblets. The entire sky was filled with the interlaced tracks.

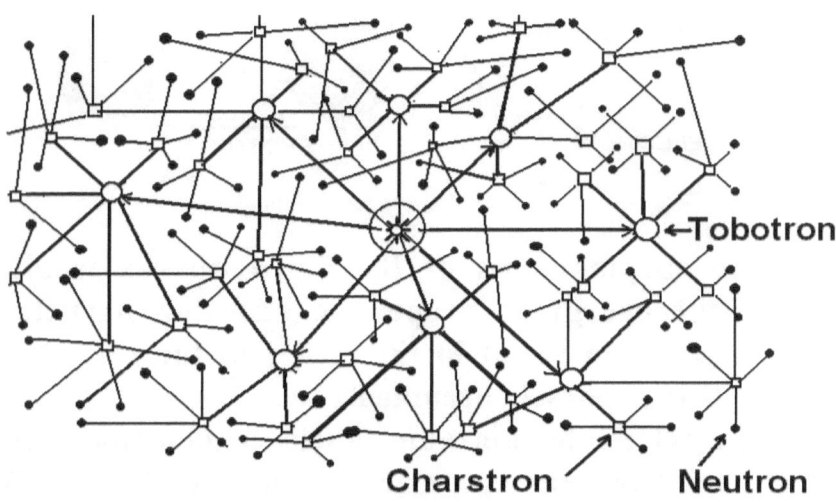

BIG BANG

In the figure above, only four or five particles are shown leaving each explosion point, instead of the proposed 100.

It can be seen that many of the final particles (neutrons) would be moving back towards the center of the big bang. I assumed that it was there that most of the neutron-to-neutron collisions occurred and produced the protons and electrons that would become the visible universe.

The figure below shows a neuron-to-neutron impact. Three of the down quarks, each with a charge of one third, would combine into an electron, with a charge of unity. The two up quarks and the remaining down quark would combine into a proton.

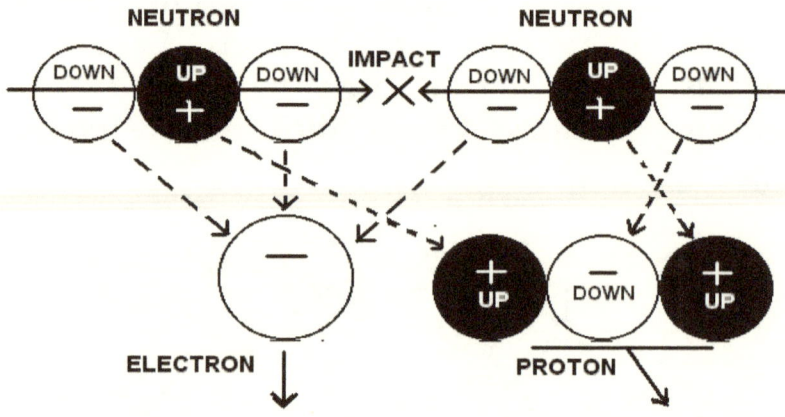

The surplus mass of the down quarks would produce the energy to propel the resulting particles away. This would result in the center of the resulting universe being sparsely occupied.

The relatively thin outer shells of neutrons and charstrons of the universe-sized black hole (Univoid?) would have been propelled outward at faster than today's velocity of metz. These would become the tenuous outer reaches of the present universe.

Most of the mass of the universe (neutrons and protons) would have combinations of radial and tangential velocities and would produce the great majority of the distribution of galaxies, stars and dark matter, occupying the middle region of the present universe.

In the previous book I raised the question as to whether there would be a sufficient number of metz to account for gravity and the speed of light, throughout the universe. To answer this question requires an estimate of the size of the visible universe and the number of metz in it.

BLACK HOLES

You can see the full chapter on black holes later in the Appendix on page 100. It seemed best to begin, in describing my concept of the evolution of black holes, by going one step backwards to neutron stars (which I renamed "neutroids" since they are not stars). My library research indicated that "exotic particles" were thought to occur at the center of neutroids. This triggered my idea that the exotic particles would be another form of neutral particle.

The idea of a solid body made almost entirely of neutrons, all in contact with one another, made me wonder if any of the metz impacting on its surface could make it through to exit the opposite side. This would seem to be necessary, for any photon impinging on the surface, to be reflected. A metz would propel it away and make the body visible. I do not think that metz can penetrate a body of solid neutrons. A possible explanation occurred to me, as suggested by the following figure.

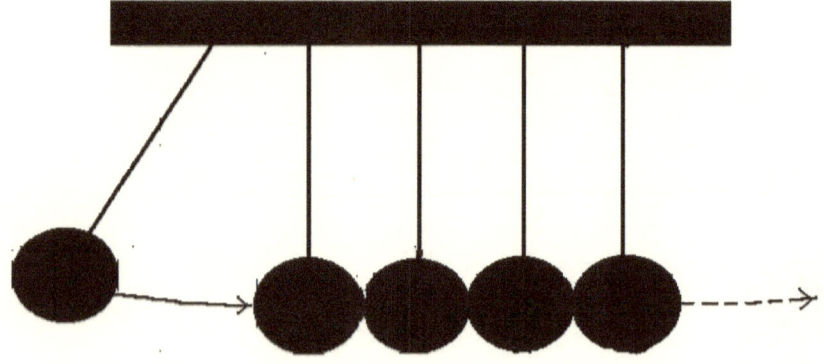

Transfer of Momentum

From the above diagram, of a common demonstration object, it seems possible that the momentum of a metz, impacting on a surface neutron, could be transmitted through the neutroid. Thus causing a different neutron to be expelled from some other point on the surface. It would then become a metz capable of propelling a photon away from the surface. This may only occur for a small percentage of the impacts. Most of the metz would dissipate their energy as heat, and in applying pressure at the center. They would also add to the mass of the neutroid.

In the following figure the possible evolution of black bodies is described. They range from the black body, from a single large star, all the way to a single body containing all the mass of the entire visible universe. These different bodies in this diagram are only suggestive of possible sizes and proportions.

BLACK HOLE EVOLUTION

LEGEND

Neutrons	(Up and Down Quarks)	☐
Charstrons	(Charm and Strange Quarks)	☐
Tobotrons	(Top and Bottom Quarks)	☐

Black Hole	Galaxy Core	Galaxy	Visible Universe
10 miles	100 miles	2000 miles	10 million miles

NOT TO SCALE

The black hole core is shown as charstrons. The size of the core would be such as to be able to sustain the neutron mass above it, without further collapse.

The galaxy core black hole is suggested to have a core of tobotrons, surrounded by a shell of charstrons and an outer shell of neutrons. As I indicated, the relative sizes of all the layers are to be determined.

A possible explanation of why these bodies are black is that light cannot reflect from them because no metz leave them to propel the photons away. The following figure suggests why this may be so.

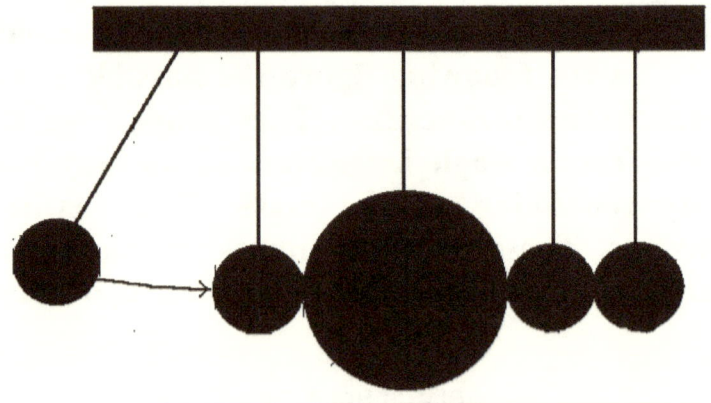

Transfer of Momentum ?

It seems unlikely that the impact of the small ball at the left would transfer enough momentum through the large ball in the center to move the small ball on the right.

Similarly, the impact of metz on the outside of the above black hole would transfer their momentum into the charstron core and not through it. The momentum of a metz impact could not get back out to the surface of the body. Photons would not be propelled away and the body would be invisible.

DARK ENERGY

I consider dark energy to be the kinetic energy possessed by all of the dark matter particles traveling at their characteristic velocity. This velocity averages the present speed of light. Reference material, that I have found, calls these particles, "wimps". This stands for "weakly interacting massive particles". I prefer the name metz instead of wimps.

The amount of visible matter is stated to be less than 4% of the total mass of the universe. The dark matter is estimated to be about 23% and the dark energy is the remaining 73%.

If one insists on using the equations of relativity, for the hypothetical rest mass of the dark matter to be increased from 23% to a relativistic mass that is 96%, four times as great, their velocity would have to be within about 3% of the speed of "light".

This was calculated using the following equation.

$M_r = M_o / [\{1-(v/c)^2\}^{1/2}] = 4M_o$

Or $1/16 = 1-(v/c)^2$

And $v/c = (15/16)^{1/2}$

Therefore $v = .968c$

A classical explanation is to consider the kinetic energy of a metz as:

$\frac{1}{2}(M_n) \times (V_c)^2 = \frac{1}{2} (1.67 \times 10^{-27}) \times (3 \times 10^8)^2$

This equals 7.5×10^{-11} joules, as the kinetic energy of a metz. The total dark energy of the visible universe would be 7.5×10^{69} joules.

NEUTRONS

These theories all revolve around the neutron. My concept of the configuration of the quarks in a neutron is based on the fact that the quarks have electric charge and electric fields. Due to their spinning charge, they will have magnetic fields. These fields should hold the quarks together. The following figure shows how the magnetic fields tend to do this.

NEUTRON

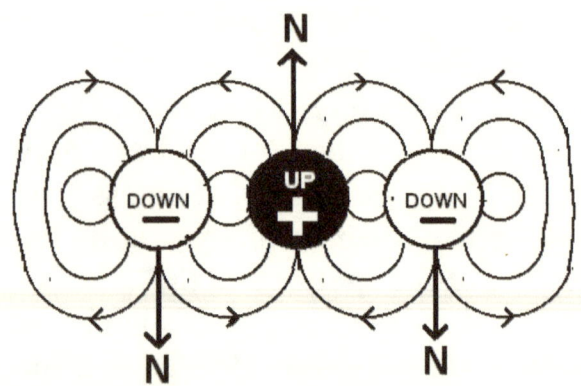

Magnetic Fields

One could lay three bar magnets side by side with the polarities arranged as those above and they will move together and hold each other firmly.

In the figure below, I present my recollection of how the electric force fields surround oppositely charged bodies. These alternate charges will also tend to hold the quarks together.

NEUTRON

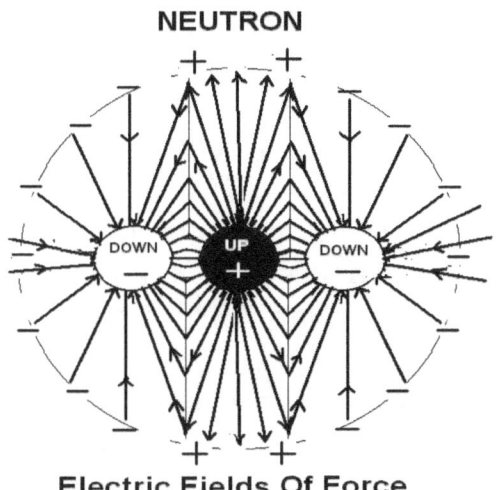

Electric Fields Of Force

Because the positively charged quark is bracketed by two negatively charged quarks, I assume that the surrounding, superimposed electric fields of the neutron would be more negative than positive around the circumference. As a neutron is racing past an object, the spinning fields would effect the object more by the longer duration, negative, force fields than by the shorter duration, positive ones. If the spin is not considered significant, the probability is greater that a negative field will interact with an object, in passing.

PROOF

Since I wrote the "highly speculative" chapter on faster-than-light travel, I have been increasingly doubtful of its possibility. This is mostly because I cannot picture sufficient numbers of head on collisions of metz occurring to produce sufficient protons and electrons to give adequate thrust to propel a spaceship.

If a stationary, experimental setup could be developed that produced these collisions, under controlled conditions, it would go a long way to verify my basic assumptions about metz and dark matter. I anticipate that the collisions would occur only occasionally. I assume that the metz would usually tend to repel each other, even when they are on an exact, head on path. It seems that some thing might be done to orient them, so they would attract each other. I show two possibilities below

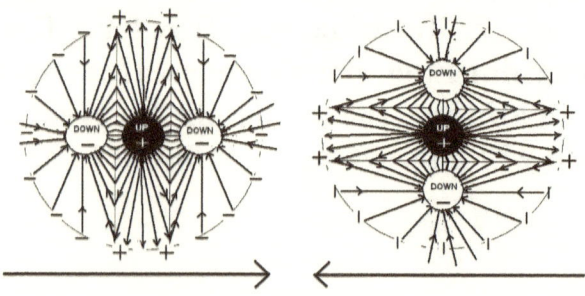

In the above configuration the maximum positive electric field of one metz would face the negative electric field of the other, just before impact. Another possibility is shown in the following figure.

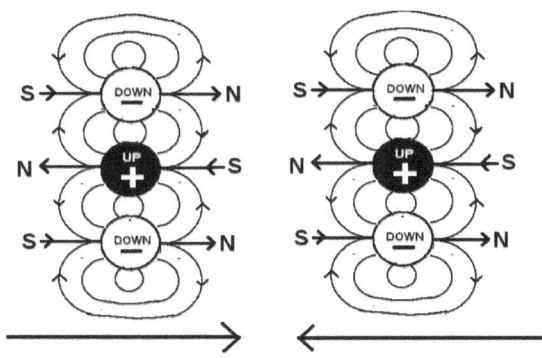

Above, the magnetic poles of the quarks in one metz, face the opposite polarity magnetic poles of the quarks in the other metz. This should produce an attractive force, between them, just before impact. The questions remain as to how to produce these final configurations and if this collision will produce a proton-electron pair. The following is an experimental setup to try.

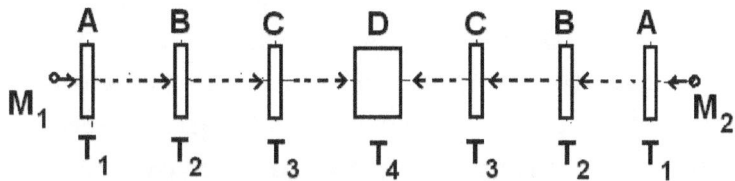

METZ EXPERIMENT

In the operation of the experimental setup, two metz, M1 and M2, will enter the "A" elements, at the opposite ends of the apparatus. This must occur at approximately the same instant, during the activation of the A elements, at time T1.

The B elements will be activated at a time T2, after a time delay that will be determined by the distance from A to B. and the time taken by the metz to travel this distance.

95

As an example, if the distance between A and B is three meters the time delay would be 10^{-8} seconds (10 nanoseconds).

The activation of the C elements will occur at T3 after the proper time delay. The element D will be a particle detector for the expected products of the metz collision. It will be activated at time T4.

The three elements on each side of the detector are only for illustration purposes. The number required might be more, but probably not less than three. These units will be designed to cause the metz to travel along the axis of the setup and arrive at the collision point with the desired orientations for maximum collision probability.

I assume that the function of these units will involve pulsed magnetic and electric fields. The pulse widths would probably be about half that of the time delay durations. The particle detector should be long enough to allow some variation in the location of the collisions. There will be some variation in the times the metz enter the apparatus, during the pulse at time T1.

The repetition rate of this procedure will also depend on the reset time of the detector and the recharging of the electronics providing the pulses. It is likely that the detection of a collision will occur only rarely. Although the number of metz coming from all directions is extremely high, the number that satisfy the conditions for this experiment will be very small.

The apertures of the devices will be limited to a size that will allow adequate pulsed field strengths to occur during the short duration pulses. A couple of possible device configuration to produce various electric fields is shown below.

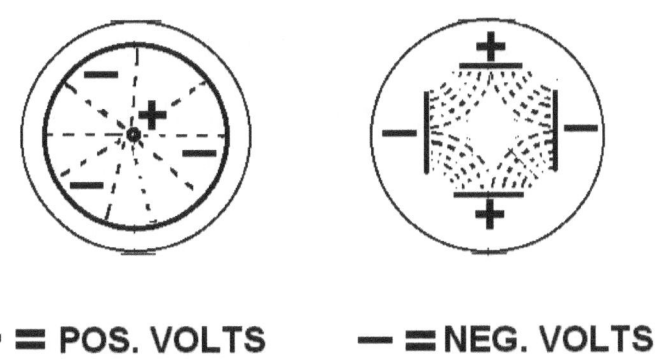

+ = POS. VOLTS   — = NEG. VOLTS

In the figure above, the diagram on the left shows a radial electric field. The small anode in the center has a positive voltage. The circular cathode surrounding it has a negative voltage. Of course these polarities could be reversed.

The diagram on the right shows a more complex electric field. Two pairs of plates are shown but there could be more than this number. Several of this type of device, placed in sequence, might control the spin of a metz.

Simple current loops could be used to generate magnetic fields concentric with the device axis. These fields might polarize the metz to the proper orientation and move them to the axis.

WEIGHT

The following figure should be considered in calculating the weight of objects in the vicinity of the Earth.

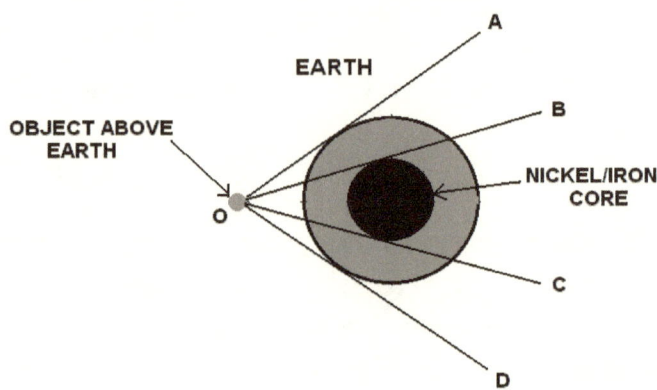

I think that the atoms, in the dense nickel-iron core of the Earth, could have many close encounters with a metz. This would result in many of the metz being deflected and transferring momentum to the core. This could account for the high pressure on the core and its high temperature.

As a first approximation, the core could be considered as blocking most of the metz coming through the solid angle, subtended by the core, at the small object.

Therefore, most of the momentum from the metz propelling the small object towards the Earth would be coming from the direction opposite to the Earth, through that same solid angle as that to the core.

It is obvious that any metz whose path would have intercepted the small object, if the metz were not deflected, would miss the object, if it were deflected. This will occur 100% of the time.

The question remaining is; how many of the metz that would have missed the small object, if the metz were not deflected, would intercept the small object if the metz were deflected? In the next figure I address that question.

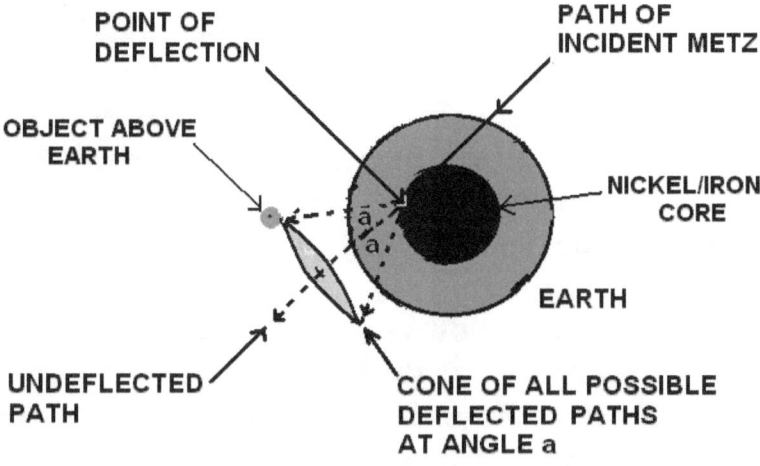

A metz is shown on a path that if it were to continue would have it missing the object. A deflection point is shown that for a specific angle a, and only one of the many possible deflections, the metz will intercept the object. Obviously the probability of this occurring is extremely low.

BLACK HOLES

Chapter from "Beyond Einstein"

Our Sun is a second-generation star. It is one of the smaller sized stars. It is used to compare the size of other stars. We call our Sun "one solar mass". The smallest stars are about one half a solar mass and the largest are more than one hundred solar masses in size.

The amount of nuclear fusion needed to withstand the weight of a star varies with the size of the star. Our Sun has enough fuel to burn for billions of years. The largest stars may have to use up all their fuel in less than a million years. When a star has used up its fuel, a number of different things can happen. Some stars become large cool red stars. Some become small hot dwarf stars. Stars that are in the range of ten solar masses can explode and then collapse, becoming a neutron star.

The most spectacular event occurs when a star of about one hundred solar masses explodes as a super nova. It may then collapse into a black hole. Mathematicians have done much of the theoretical work on black holes. Their equations for the characteristics of matter at the end of the collapse of a star into a black hole, predict that the size of the body goes to zero and the density becomes infinite. However, these bodies are said to have a finite mass. This is called a singularity. I think this concept is hard to believe. I prefer one that is simpler and more logical.

The nature of neutron stars may be the best place to start in understanding my idea about black holes. Neutron stars are thought to have a surface shell of iron atoms, which are the remnants of the depleted nuclear core.

Inside of the surface shell are the neutrons, essentially in contact with one another. In the collapse of the core the electrons of the heavier atoms are forced into the nuclei. The electrons and protons combine, forming neutrons.

It is theorized that at the center of the neutron star some exotic particles could be forming. I wondered what these exotic particles could be. This led to the idea of neutral particles, made up the next heavier quarks, becoming the predominant component at the center of the individual black hole from a collapsed star. This particle should be a hundred times more massive than a neutron.

The black hole should still have a layered structure with an outside layer of iron. Within that there should be a layer of neutrons and inside of that there should be the particles made up of the charm and strange quarks. I really need a shorter name for those particles, such as "charstrons", perhaps.

Neutron stars are said to result from the collapse of stars of about eight solar masses. After they collapsed, they had masses of a little over one solar mass. I am assuming that this was essentially the mass of their nuclear core. The remainder of their original mass must have been mostly hydrogen gas blown off in the super nova. I understand that these bodies ended up with a diameter of about ten miles.

A scientist named Schwartzchild derived an expression that gave the relationship between a body's mass and the distance from which light could no longer escape.

For a one solar mass neutron star with a diameter of about ten miles, the Schwartzchild limit is a four-mile diameter sphere. This would be within the body of the neutron star. Therefore, the neutron star should still be visible.

In the following figure the relative size of different mass stars is shown. Their masses increase with their volumes, and their volumes increase as a function of their diameters cubed.

STAR SOLAR MASSES

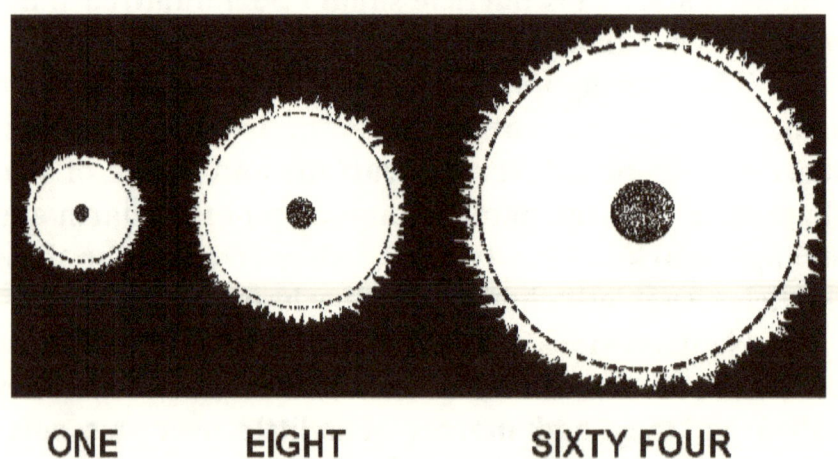

ONE   EIGHT   SIXTY FOUR

Black holes are supposed to occur some of the time when larger stars, in the range of fifty to one hundred solar masses, collapse in a super nova. I am assuming that a sixty-four solar mass star would have a core of about eight solar masses. If it had collapsed into a neutron star, which it might in a transient phase, it would have had a diameter twice that of the above neutron star. That is about a twenty-mile diameter. It should still have about the same mass of eight solar masses.

For this mass the Schwartzchild limit would be a sphere with about a thirty-two mile diameter. Therefore if this neutron star remained a neutron star it would be close to being a black hole.

However, assume that many of the neutrons in the center of this body were converted to charstrons. If the ratio, of neutrons being converted to charstrons, were about a hundred to one, the twenty-mile diameter of the eight solar mass body, would be reduced by a factor of about four. That is assuming the individual charstrons were about the same size as neutrons. It would stop collapsing at about a diameter of five miles. This would definitely be a black hole.

In the above examples, I have assumed that the Schwartzchild equation is valid. Since it is partially based on an Einstein theory it may not be.

If this first kind of black hole were in a situation where it was absorbing more and more matter from adjacent stars, it might reach a state where it would collapse again. A large number of "charm" and "strange" quarks would be converted into a smaller number of the more massive "top" and "bottom" quarks.

These quarks would become neutral particles about one hundred times as massive as the charstron particles. Perhaps, we could call them "tobotrons". They might reduce the diameter of a charstron black body by a factor of four in becoming an equivalent tobotron black body.

I am assuming that it would be configured with a layer of iron on the outside. Within that there should be a layers of neutrons and charstrons.. The tobotrons would form within the layer of charstrons.

Perhaps a different name, other than black holes, would be appropriate. Astronomical bodies are commonly named using the suffix, "oid". That is like asteroid, meteoroid and planetoid. I would like to suggest that neutron stars be called "neutroids", charstron black holes be called "charstroids" and tobotron black holes be called "tobotroids".

The particles that would result from the conversion of quarks of one type to those of another are shown in the following figure. There is the possibility that the charstron quark charges are 100 times those of the individual neutron quarks. Also, this would imply that the charges of the tobotron quarks could be 100 times that of the charstron quarks and 10,000 times that of the neutron quarks.

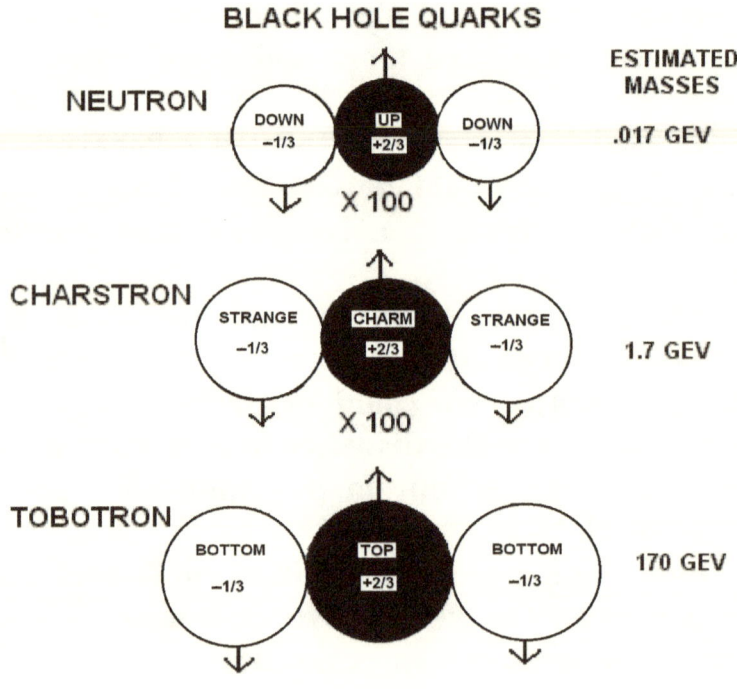

It is interesting to note that the above tobotron has about the same energy equivalent mass as that attributed to the Higgs particle, 170 GEV.

There is evidence that there are massive black holes at the center of most galaxies. They appear to have masses in the order of millions of times that of our Sun. This suggests to me that these could be black holes of the second kind (tobotroids). If a hypothetical tobotroid of ten solar masses would have a diameter of about one mile, a ten-million solar mass tobotroid, in the center of the galaxy, would have a diameter of about one hundred miles.

Assume that there are a hundred billion solar masses of stars in the galaxy. If all were ingested into this central black hole it would have a diameter increased by a factor of the cube root of 10 thousand, which is about 22. The final diameter of the galaxy weight black body would be about 2200 miles. (We could call this a galaxoid.)

If a hundred billion such galaxoids combined into a single massive black hole, its diameter would have increased by a factor of the cube root of a hundred billion (about 4500). Therefore the final diameter of a universe weight black body (univoid?) would be about ten million miles. This supposedly accounts for the present visible universe. Since most of the gravitons will remain dark, assume that they represent 7/8 of the total volume. This suggests a final diameter of twenty million miles.

It is conceivable that even more massive quarks could form at the center of such a body. These could create another type of neutral particles out of tobotrons. This could reduce the diameter of the universal black body even further.

Consider how black holes will interact with other matter. Gravitons will impact on a black hole from all directions. The gravitons will enter through the spaces between the surface atoms and make innumerable elastic collisions within, transmitting their momentum into the black hole.

Any material body in the vicinity would be accelerated toward the black hole, since no gravitons would be coming out of the black hole to impede them. The approaching bodies would achieve exceedingly high velocities and be assimilated into the black hole upon impact.

When black holes are in the same vicinity, they will tend to shield each other from gravitons coming from the directions beyond them. These gravitons will impart momentum to the black holes they fall upon and cause the black holes to move towards each other, until they merge.

It is evident that black holes would accumulate gravitons. These gravitons will have come from their own universe, from adjacent universes, and even from previous universes that no longer exist. The gravitons will be recycled, liberated when the black holes containing them turn into a Big Bang.

(This is from a self-portrait by the author)
Circa 1967

Lenard Metzger was born and raised in Rochester, New York. He has lived there ever since, except for the years spent in service and college. He received a degree in physics from Ohio State University.

He was a long time employee of Eastman Kodak Company. After retiring he returned to his early interests in art, science and writing.

www.ingramcontent.com/pod-product-compliance
Lightning Source LLC
Chambersburg PA
CBHW022101170526
45157CB00004B/1432